Debriefs and Postmortems for Product Development

Your illustrated guide to improving performance through lessons learned
4th Edition

By: José Campos

Published by:

Debriefs and Postmortems for Product Development (4th Edition)

By: José Campos

Published by:
Rapidinnovation Press
Rapidinnovation, LLC
25000 SW Big Fir Rd.
West Linn, Oregon 97068
United States
www.rapidinnovation.com

All rights reserved. No part of this book may be reproduced or transmitted in any form or by any means, electronic or mechanical, including photocopying, recording or by any information storage and retrieval system, without written permission from the publisher or authors, except for the inclusion of brief quotations in a review.

Copyright ©2009-2012 by Rapidinnovation Press The content of this book is the intellectual property of Rapidinnovation, LLC, which has granted License to Rapidinnovation Press to publish this content.

ISBN-13: 978-0-9817595-1-7
ISBN-10: 0981759513

Published in the United States

CIP data available from the publisher

Get your interactive PDF version FREE!
See the last pages of this guidebook for details

4th Edition

Introduction

Introduction
- Legend 6
- Approach 7
- About the Author 9

Postmortems
Critical Definitions and Concepts
- Introduction 11
- Timing of Debriefs 15
- Possible Focuses for Debriefs 17
- Four Critical Decisions 19
- Example: Selecting the Right Project to Debrief 20

Conducting Debriefs
- Key Questions to Consider 22
- Getting to the Root of Things 23
- Examples: Effectively Describe Issues and Opportunities 24
- Rules and Strategies for Debriefs 25
- High Level Process: Debriefs 26
- Detailed Process: Debriefs 26
- Create a Well-Documented Action Plan from a Debrief 32
- Example: A Well-Documented Action Plan 32
- Example: Debrief Planning Tool 33

Deployment and Long-Lasting Changes
- How to Actualize Productivity Improvement 34
- Spreading the Knowledge 35
- Example: Debrief Planning Tool by Focus (Topic) 36

Critical Tools
Deployment of a New Tool
- Cascading a New Tool to the Organization 40
- Detailed Process: Deploying a New Tool 42
- Planning Checklist 44
- Planning Document 45
- Tips for Successful Deployment 46

Yellow Sticky Protocol
- Effective Group Participation 47
- Level 3 Detailed Process: Yellow Sticky Protocol 48

Introduction

    Example: Yellow Sticky Protocol 49
    Examples: From Requirements to Features and Attributes 50

Affinity Diagrams
    Organize Information and Reach Consensus. 53
    High-Level Process. 54
    Detailed Process: Affinity Diagrams 55

Prioritizing Affinity Diagrams
    The Process to Prioritize 57
    Level 3 Detailed Process: Prioritizing Affinity Diagrams 58
    Example: Affinity Diagram 59

Problem-Solving Process
    Problem Solving Process 60
    Detailed Process: Problem Solving 61
    Problem Solving 64

Karnaugh Maps: 2x2 Matrixes
    Prioritize Items from a Selection of Many 65
    Example: Karnaugh Map (2x2 Matrix) 67
    Example 2: Prioritizing Customer Requirements 68
    Pair-Wise Comparison 69
    Detailed Process: Prioritizing a Karnaugh Map. 70

The 5 Whys
    The 5 Whys 72
    Examples: The 5 Whys. 73

The Fishbone Diagram
    The Fishbone Diagram 75
    Detailed Process: Fishbone Cause and Effect. 77
    Developing Scales for Prioritization 80

Process Mapping
    Steps to Success 82

Appendixes
Appendix A: Recommended Books
    Introduction. 84
    Teams and Teamwork for Product Development 84
    General Management and Strategy 85
    Product Development. 85
    Productivity Improvement. 86

Introduction

4th Edition

    Project and Program Management 87
    The Voice of the Customer (VOC) 88
    Risk Management and FMEA (FMCA) 90

Appendix B: Glossary
    Definitions for Product Development 91

Appendix C: Templates
    Instructions 114
    Selecting the Right Project to Debrief 115
    A Well-Documented Action Plan from a Debrief 116
    Debrief Planning Tool 117
    Debrief Planning Tool by Focus (Topic) 118
    Planning Checklist for Deploying a New Tool 120
    Planning Document for Deploying a New Tool 121

Get Your Interactive Version FREE! . . . . *details in the back*

Other Titles You May Be Interested in Owning . . . . *details in the back*

For comments, questions or additional copies please e-mail the author: debriefs@rapidinnovation.com.

Debriefs and Postmortems for Product Development

Introduction

Legend

Navigate through this Guidebook

To ease navigation, we have provided a set of icons, which provide orientation and clarity. They are consistent throughout the guidebook, always carrying the same meaning.

Symbols to Indicate Facts to Know

Important Suggestion
Used with each important suggestion, note, or tip, derived from experience in the field during actual projects.

Essential Quote
A quick way to read very important information. We suggest you read all essential quotes.

Flexibility
Indicates the flexibility of a particular tool. We encourage you to adapt the tool to your specific situation and not be governed by rigid, prescriptive frameworks.

Definition
A defined term. See also our Glossary at the back of the guidebook.

Time Frame
To help you plan the likely duration of a meeting, work session, or other event.

You Are Here
One more way to get oriented or to check that you have take the proper steps up to this point.

Symbols for Links to Helpful Info

Note for Print Version
Interactive features are available in the electronic PDF version. If you have a printed version, contact the authors to learn how to upgrade your guidebook.

Template (Printed Version)
You will find copies of the templates at the back of your guidebook.

Template (Electronic Version)
Double click this icon to open a corresponding template (enabled for use in Adobe Reader and Acrobat) that you can fill with your own information and share with your peers and colleagues. When you close the template, you will return to this guidebook.

Each time you open a template from your guidebook, it will be fresh. But you can save versions that you have completed, partially filled in or left blank for future access. Saved versions remain enabled for revisions and collaboration.

These enabled templates also provide you with Comment & Mark-up Tools. Just as you might with a hard copy, you and your team can highlight text, add notes and make custom annotations.

See Also (Electronic Version)
Clicking this icon will jump you to a place in the guidebook with more information relevant to the topic at hand.

Back Button (Electronic Version)
After following a link, you can easily click to return to your original place in the guidebook.

For comments, questions or additional copies please e-mail the author: debriefs@rapidinnovation.com

Introduction

Approach

Why We Wrote this Guidebook

Debriefs have been shown to be the best way to improve the productivity of an organization — even better than training. We set about creating this guidebook to provide product development teams (R&D) with the tools, templates, examples and inspiration to use debriefs to improve performance or, more specifically, reduce time-to-market.

Product development (R&D) is a tumultuous world where it's extremely important to know what works and what doesn't. Debriefs, properly executed, will methodically help you eliminate waste and identify best-practices that improve productivity.

A How-To Manual

Having been on the receiving end of guidebooks, handbooks, "primers" and training during our lives in the corporate world, we realized that often the information was not sufficient for us to actually use it in our situation, or the information was formatted poorly such that it made it impossible for us to apply it, which resulted in much frustration, not to mention loss of productivity. Worse yet, the documentation we received from training sessions was simply a hard copy of the PowerPoint slides shown during the session, which were of little help to us when it came time to apply the learning.

We made a commitment that our guidebook would contain everything a professional would need to apply every concept in it. We have strived to provide the tools, methods and encouragement for you to successfully understand and apply the voice of your customers.

How to Use this Guidebook

We developed this guide for individuals and teams who are in product development (Marketing and R&D) and have specific assignments to deliver results. The entire focus of the guidebook is to lead you — step-by-step — through each and every concept, even if you have never performed the task before.

The sequence of this guidebook is based on a chronological approach — the outline and format follows a logical "starting point" and move towards a "end point." This means that the higher the page number, the later in the process.

This is not a book about concepts. We have limited concepts to the minimum necessary to create context or perspective. The focus of this guidebook is on the "how-to."

The Process Maps

The most important element of this guide is the process maps — the step-by-step directions for starting and completing any of the key tasks, methods and templates in this guidebook.

These process maps will help you implement any of the tools. They have been developed over many years of actual application and experience.

"The most important element of this guidebook is the process maps — the step-by-step notes for starting and completing key tasks."

The Template and Tools

Templates are forms that contain information that you and your team enter in order to collaborate and create archival information for later use. Tools are matrices, spreadsheets, tables and other instruments.

It is important to note that for every tool or template there is at least one example and in some cases several. This is intended to provide you with a general idea of the information that should be entered. It is not intended to limit what you and your team feel needs to be done.

Introduction

Unique Capability for the Electronic Version
Not only does every template in this book have a full example, but also it is fillable. You and your team can enter information about your project and share it. You will notice a link in the form of an icon and when you click on it, you will be able to open a blank template that you and your team can use. Once you have completed entering the information, the template becomes its own PDF independent of the guidebook. You can save it using your own name or title, which you can also share with the rest of your team around the world. You can continue entering information until you are satisfied.

Flexibility
We understand that every development project is different — by definition. Consequently, our intent is not to limit your flexibility. All the process maps, templates and tools can be modified to better meet your objectives and style. We suggest that the first time you apply the these process maps, templates and tools that you try to adhere to our format. As you become familiar with the intent, you can modify them until you have something that is unique to your organization and also meets your requirements to the fullest. Don't be afraid to experiment and modify.

Mission of this Guidebook
The mission is to help you and your team improve the productivity, efficiency and profitability of your organization through the use of debriefs.

Objectives of this Guidebook
The overall objective is to create the capacity to improve future performance! And the results of using this guide will include these specific objectives:

- You will be familiar with overall strategies needed to establish debriefs as part of your New Product Development (NPD) process.
- You will be able to apply the material presented in this guidebook to real-world situations.

Your Input
No book is perfect; this one is no exception. If you find difficulties in using the tools, concepts and processes, do not hesitate to contact the authors: debriefs@rapidinnovation.com.

4th Edition

Introduction

About the Author

José Campos, Principal Consultant and Founder, Rapidinnovation

José collaborates with B2B technology companies around the world on the methods, tools and cultural shifts to rapidly and consistently deliver superior value to customers.

José has spent his career working to define and deliver innovative products to market. For 30 years he has passionately focused on discovering new methods and approaches to discover and deliver superior solutions to the customer in the tumultuous world of high-tech

He has written several acclaimed books: *Voice of the Customer for Product Development* — a methodology to discover, prioritize, and use customer requirements to create clearly differentiated products — faster than the competition. Also, *Risk Management for Product Development* — the strategies and methods to reduce the impact of uncertainty on your investment. Jose was coauthor of Project Management Toolbox in collaboration with Dr. Dragan Milosevic.

His focus on high-tech and his technical background allow him to provide meaningful and practical coaching to technology professionals. He has traveled the world help R&D and Marketing teams improve their ability to create innovative products that customers love.

Related Guidebooks for Product Development

In addition to *Debriefs and Postmortems for Product Development*, our library includes several related guidebooks in the Product Development series:

- The Voice of the Customer for Product Development
  Your illustrated guide to obtaining, prioritizing and using customer requirements and creating winning products.
- Actionable Metrics for Product Development
  Your illustrated guide to developing and using metrics to improve product margins and reduce time-to-market.
- Flexible Project Management for Product Development
  Your illustrated guide to making project management work in fast-changing environments.
- Risk Management and FMEA for Product Development
  Your illustrated guide to reducing time-to-market through risk management and FMEA.

Training, Consulting and Facilitation

Our highly experienced authors are also available to provide direct consulting to your organization. They can be team facilitators to help you implement the contents of these guidebooks. They can provide training on the topic plus many other areas of product development. Contact us for more information.

For comments, questions or additional copies please e-mail us: jose@rapidinnovation.com.
Visit us at www.rapidinnovation.com.

Get Your Interactive PDF Version FREE!

See the last pages of this guidebook for details on how to acquire your free copy of the interactive PDF version.

Postmortems

PRODUCT DEVELOPMENT

Critical Definitions and Concepts 11
Conducting Debriefs 22
Deployment and Long-Lasting Changes 34

Product Development

Critical Definitions and Concepts

Introduction

First, we must clarify the name of this methodology. It goes by several names depending on the industry and function you represent. What we will refer to throughout this guidebook as "Debriefs" is also referred to as Postmortems, Postmortem Reviews, Retrospects, Retrospectives, Lessons Learned, After Action Report and Post-Project Reviews. They all have the same purpose: Obtain information from projects in order to improve the future productivity of the organization. We recommend that you use whatever name you have used in the past and forget about the others to avoid unnecessary distraction.

Studies have shown that the best way to improve the productivity of any organization is through a robust program of debriefs. They allow the team to capture real information from actual experience. This information is critical to improve the performance of future projects.

Debriefs are well-orchestrated activities that capture targeted information from a development team for later use to improve the organization's overall performance. This information can be lessons learned, key learnings, process improvement and any other data or insight that can be used to reach higher levels of excellence.

Debriefs by Many Other Names

What we will refer to throughout this guidebook as 'Debriefs' is also referred to by other names. They all have the same purpose: Obtain information from projects in order to improve the future productivity of the organization.
- Debriefs
- Postmortems.
- Postmortem Reviews
- Retrospects
- Retrospectives.
- Lessons Learned.
- After Action Report.
- Post Project Reviews.

NEW TOPIC

Definition

- A debrief is all about creating the capacity for action to improve future performance.
- The purpose of a debrief is to create learning for the individuals in the organization, to sustain positive improvements and to improve future performance. It is about a search for knowledge and opportunities.

A debrief is all about creating the capacity for action to improve future performance.

Strategic Intent

The intent of every debrief is to answer several critical questions about a specific project. Every debrief must conclude with a plan of action: What must we do differently next time?

Strategic Questions That Debriefs Answer	Tactical Questions That Debriefs Answer
♦ What was supposed to happen? ♦ What actually happened? ♦ Why were there differences? ♦ What must we do differently next time	♦ What worked? ♦ What didn't? ♦ Why? ♦ What must we do differently next time?
This set of questions explores deviation from the plan. Deviation, while not inherently wrong — particularly in the tumultuous environment of product development — may indicate an opportunity for improvement. The process isn't complete until you understand "why" it happened.	This version is more tactical. The focus is still on deviation from the plan, unanticipated results and the root cause of the change. But these questions also suggest what changes to future behaviors need to be made to produce a desirable outcome next time.

For comments, questions or additional copies please e-mail the author: debriefs@rapidinnovation.com.

Critical Definitions and Concepts

Key Assumptions

For debriefs to work on a continuous basis, the following assumptions must be true:

- The information that was collected and prioritized is put to use to improve specific processes, procedures or tasks.
- Debriefs are an on-going, enduring undertaking for your organization, not an occasional, isolated activity.
- There is a documented process to deploy the information and to incorporate it into the practices of your organization, which will result in measurable performance improvement.
- There is a small and simple set of metrics to track progress.
- Debriefs are in actuality and perception non-punitive and guilt-free.

Debriefs have been shown to be the best way to improve the performance of an organization.

The Business Case

In the words of the famous Spanish philosopher George Santayana: "Those who cannot learn from history are doomed to repeat it." Organizations cannot afford to keep repeating defects and mistakes due to their impact on the bottom line. They also interfere with your ability to compete in an ever more complex world. Neither can organizations overlook innovation nor opportunities for improvement derived from well-orchestrated debriefs.

Product development is a chaotic environment where constant change is the rule. Further, product development is becoming more complex — for example, multi-site product development, strategic partnerships, co-development and much more. All of this is uncharted territory, not the ordinary process of years past. This new environment requires constant examination though a robust framework of debriefs and their corresponding improvement plans.

Finally, over the years, we have validated that debriefs are a fine tool for team building. They improve morale, re-energize the development teams and foster a collaborative environment. This is due to the consolation that problems will be addressed and good practices will be repeated. Debriefs create a vivid environment of continuous improvement and the reassurance to your development team that things will continually improve.

We estimate that a well-orchestrated debrief can improve productivity by one percent. Four debriefs per year is our recommendation. Just imagine, in one year the productivity of your development team could be up by four percent! Now imagine conducting debreifs for five consecutive years — you could improve organizational performance in the range of 20 percent!

Product Development

Critical Definitions and Concepts

Benefits

The diagram below illustrates some of the benefits of debriefs. The list is not complete as the benefits are many. It has been shown that conducting debriefs is the best way to improve the productivity of an organization. Indeed, studies show that debriefs are better than training, meetings and other forms of education.

Debriefs should be part of the continuous improvement strategy of the organizations (Agile, Kaizen, ISO 9000, Six Sigma, etc.). In addition, they present opportunities for recognition, celebration, collaboration and support that can build stronger teams and cohesion in the organization. Finally, debriefs foster the habit of learning from experience, which helps create a learning organization.

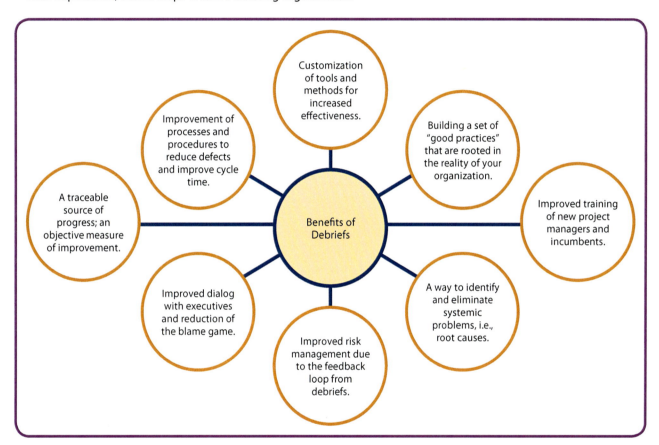

The Initial Questions

Before you call a meeting to debrief a project, there are three initial questions that must be answered:

1. Why are we debriefing this particular project?

    In order to justify the investment of time and effort, you must have a strategic reason to debrief a particular project — a criteria that, if met, flags a project as worth debriefing. We feel that not all projects need to be debriefed; rather, careful selection of the very few critical ones is more productive.

2. When will we do the debrief? (At what time during or after the project?)

    The timing of the debrief is critical so it can yield the right type of information. Debriefs are not exclusively done at the end of a project.

3. What will be the focus — the topic or emphasis — of the debrief?

Critical Definitions and Concepts

The answer to this questions is connected to first, the reason for doing the debrief. You need not debrief the entire project.

The Sad Truths

1. Most companies spend considerable resources planning and executing very expensive projects, and little or nothing on evaluating the results or learning to improve their performance, so they continue making the same mistakes.
2. Few companies examine their completed projects in any depth.
3. When they do perform debriefs, they are superficial and not documented.
4. Employees participating in debriefs fear that there are hidden agendas and other nefarious intents on the part of the company.
5. Few companies actually take action as a result of debriefs. There is no framework to ensure a closed-loop process.

This is not to discourage you from undertaking debriefs; rather, this is to bolster your resolve to ensure your debriefs are successful. Debriefs are successful when they result is visible productivity improvement.

You can take specific steps to remedy the five items listed above. Work with you team and your managers to ensure measurable results.

Resistance to Debriefs

In our research and experience we have found that organizations perceive debriefs as a very positive activity. Which begs the question: Why do so few organizations actually perform debriefs? We have found three critical reasons that keep organizations from undertaking debriefs.

Important: Not all development projects need to be debriefed. We suggest to select the few that will provide maximum leverage to improve productivity

The intent of this section is to alert you to the possible obstacles so you can take the appropriate corrective action.

1. Fear of Punishment or Retribution

Regrettably, some debriefs deteriorate into finger-pointing and other forms of complaining. Worse yet, in rare cases the information is used to punish employees. This is wrong and must not take place. It stifles participation and ruins morale.

What to Do:

- ❖ When you schedule a debrief you must reassure everyone that no one will be punished or in any way scolded for speaking freely. This is called creating a safe environment. Obviously, you must adhere to this commitment. It is the right thing to do, and it will build your credibility for future debriefs.
- ❖ Ensure that you talk to all managers and supervisors, and obtain their commitment that they will only use the information for improving the productivity of the organization and not as a tool for punishment.
- ❖ During the meeting, allow a period of time for complaining and whining. It is human. It allows people to vent. Once this is done, you need to refocus the team on the task at hand — i.e., the debrief.
- ❖ During the meeting, remind the team members that it is safe to speak freely and that no negative action will take place. Then live up to your promise!
- ❖ In rare cases, you might want to simply interview individual team members for their input. For example, two of you take the responsibility of interviewing the members of the team and writing a final report with findings and recommendations. An

Important: In product development very few projects go as planned; consequently, the need and value of debriefs.

Product Development

Critical Definitions and Concepts

alternative is to hire an outside facilitator or management consultant to perform the interviews and write the action plan.

❖ Another alternative is to hire an outside facilitator to help you with the debrief. The advantages are that the facilitator has the time to organize and ensure all preparations are done and someone neutral from outside the organization will facilitate the meeting.

2. Past Failures of Similar Efforts

This happens when previous debriefs or lessons-learned meetings did not produce any visible results. The organization is jaded or skeptical about the probability of success.

What to Do:

❖ Ensure that you are committed to take action on the input from the debrief. If you're not, then don't have the debrief!

❖ Develop an action plan with detailed information about what needs to be done, who is going to do it and when it will be done. You must exercise your leadership by tracking the implementation of the action plan.

❖ Ensure that managers and supervisors will allow employees to take some time to complete the action items from the plan.

❖ Communicate to all team members every time there are visible results from the action plan. This will encourage them to participate in future meetings.

3. Too Busy Working on the Next Project

In most organizations there is urgency to start on the next project, thus limiting the availability of people to complete the action items from the plan let alone attend the debrief meeting.

What to Do:

❖ From the start of your project, build in ample time for the debriefs. That is, include the debrief as part of the scope of the project and, more importantly, in the schedule and resource plan.

❖ Ensure that managers and supervisors will allow employees to take some time to complete the action items from the plan.

❖ Do not debrief every project — rather, carefully select the few that are likely to yield the most benefit.

Timing of Debriefs

Traditionally, debriefs are thought to be done at the end of a development project, but we feel there are several other opportunities to initiate them. We strongly suggest that you don't think of debriefs as something you do only at the end.

The timing for debriefs is critical to obtaining the maximum benefits. You need to use your judgment and decide the appropriate time and reason for debriefs. Remember, you do not debrief every development project — you must carefully select your targets.

The timing for your debriefs varies depending on your objectives. Why do you need the information? When do you need the information? And what are you going to do with it?

Critical Definitions and Concepts

Postmortems

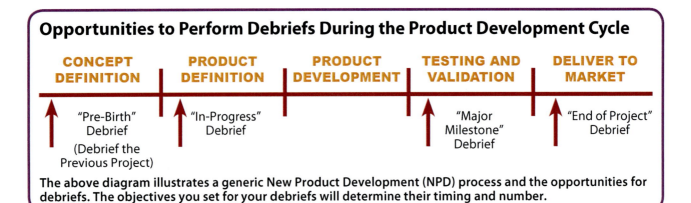

The above diagram illustrates a generic New Product Development (NPD) process and the opportunities for debriefs. The objectives you set for your debriefs will determine their timing and number.

Detailed Timeline of Opportunities

Type of Debrief	Reason	Timing of Debrief
1. Pre-Birth	• Incorporate learning from previous projects. • Team building.	• Before or at the time of your project kick-off.
2. Interim or Progress Debrief *During key dates in the life of your project such as major milestones.*	• Impact change throughout the life of the project. • Early corrective action. • Improve communication.	• Planned times during the life of your project — use your judgment.
3. Team Handoff *Your project is transferred to another organization or location.*	• Inform new team coming on board. • Formalize the handoff.	• Major milestones. • When transferring to an outsource resource. • When transferring to manufacturing. • When transferring off-shore.
4. Customer Engagement. *Critical points in your project when considerable customer involvement is required such as in contract manufacturing or custom software.*	• Early feedback from your customers on a new product.	• During development. • After the new product has been announced. • As part of the acceptance process.
5. Crisis Recovery *Very useful to objectively diagnose the reason for failure and determine corrective action.*	• Solve current problem. • Recover from a major setback. • Eliminate repeating the same errors in the future.	• When the stuff hits the fan! • As soon as the crisis is identified. • At the end of the crisis period.
6. End of Project	• Evaluate the entire project. • Closure for the team. • Benefit future projects. • Inform management of your accomplishments.	• At the end of your project. • Shortly after the end of your project.

For comments, questions or additional copies please e-mail the author: debriefs@rapidinnovation.com.

Product Development

Critical Definitions and Concepts

Possible Focuses for Debriefs

The following list illustrates some of the topics you might want to focus on. It's not a comprehensive list. You and your team can define additional focusses that are of special interest to you. Collaborate with your team to find the focusses that are critical to your future success.

See Also **Debrief Planning Tool by Focus**

Generally, debriefs are done at the project level, where the entire project is evaluated. While this is acceptable, it is more productive to focus your debrief on a specific area. Naturally, you invite individuals who can make a contribution to the chosen focus.

Focus (Topic)	Examples
1. Communications	• The effectiveness of communication within your team. • Also, with distant suppliers or contractors. • Communication with executives and managers. • Communication with your customers.
2. Planning	• The effectiveness of your planning efforts, tools and methods. • The collaboration and communication during the planning process. • Usefulness of your WBS. • Clarity of objectives.
3. Scheduling	• Focus on the duration of events, calendars. • Gannt charts. • Communication of the scheduling to all stakeholders. • Slippage and missed milestones.
4. Design Process	• The product design process. • Requirements management. • Tools and methods. • Interaction with vendors, contractors, consultants and suppliers. • Design review process. • Engineering change notices process.
5. Resources (Human)	• Conflicts, obstacles. • Number of people in the project (resource allocation). • Skills, knowledge and abilities. • Intercultural communication and diversity. • Celebrations, recognition and ceremonies.
6. Leadership and Management	• Effectiveness of leadership. • Ability to manage the projects or the processes within the project. • Dealing with stress and pressure. • Accountability and consequences.
7. Training and Coaching	• Effectiveness of the training associated with the project. • Overall training and competence building. • Coaching and mentoring.

CONTINUED

For comments, questions or additional copies please e-mail the author: debriefs@rapidinnovation.com.

Critical Definitions and Concepts

Focus (Topic)	Examples
8. Execution	• Ability to get things done when they are due. • Obstacles and barriers to get things done. • Dealing with breakdowns.
9. Teamwork and Collaboration	• Communication. • Recognition and celebrations. • Ability to communicate the common purpose. • Commitment to the success of the project and the team.
10. Reaction to Crisis	• Corrective action in view of a breakdown. • Collaboration to address an issue. • Avoidance of guilt and blame.
11. Risk Management	• Schedule accuracy due to proper risk management. • Ability to plan and monitor risks. • Dealing with risks that become reality.
12. Budgets and Expenditures	• Planning. • Accuracy of budgets. • Fiduciary responsibility with vendors, suppliers, contractors and consultants.
13. Forecasting Accuracy	• Ability to develop accurate schedules. • Forecasting budgets, cost overruns.
14. Roles and Responsibilities	• Clarity of roles and responsibilities. • Clear communication when changes were made, • Clarity of roles and responsibilities with vendors and suppliers.
15. Project Closure	• Effectiveness of debriefs. • Residual design issues. • Documentation management. • Teamwork
16. Interaction with Customers	• Customer satisfaction. • Communication with customers. • Clarity of requirements and requirements process. • Customer complaints.
17. Vendors Contractors and Consultants	• Roles and responsibilities. • Costs and estimate accuracy. • Teamwork. • Obstacles, complaints and barriers.
18. Supply Chain	• Communication effectiveness. • Financial results. • Speed. • Defects.

Product Development

Critical Definitions and Concepts

Additional Information on Focusses

Some organizations select a set of standard focusses to debrief. This is acceptable, although there is a risk of repeating to the point where no additional significant information can be obtained. This is known as the point of diminishing return. Be careful not to fall into the trap of creating a formality or a perfunctory exercise with little or no value.

Your debrief strategy must be dynamic and able to change as your environment changes. Keep in mind also that every development project is different, so it makes little sense to become locked into a single approach for debriefs.

The same can be said about checklists, which can be useful, but can also lull you into a routine with little new information or insights. Checklists are used to ensure that all appropriate focusses are addressed. The risk is that after a few times it may become simply a way to pass the time. The intent of a debrief is to dialog, discover and address issues and opportunities, not to read a list.

Four Critical Decisions

There are four specific challenges to consider when planning, running and following up on a debrief. Work with your team and develop strategies and tactics to address them.

1. **WHICH PROJECT:** Which project to debrief?
2. **WHICH IMPROVEMENTS:** Which specific improvements will you implement?
3. **HOW TO DEPLOY:** How will you effectively deploy the needed changes to the organization?
4. **HOW TO SUSTAIN:** What is your strategy to make the needed changes a permanent part of the culture of your organization?

First Critical Decision: Which Project to Debrief?

The first challenge is to decide which projects you should debrief. Quality trumps quantity! Your organization has limited capacity to assimilate new information. So it's best to carefully select which improvements you will actually deploy.

We suggest that you start with a small number of projects per year — perhaps three to four debriefs per year or 10 to 20 percent of your projects. If in doubt as to how many, try two to three per year to start, then increase the number based on your ability to implement innovations resulting from them. It's best to do fewer due to variables such as the workload of your team or organization, the nature of your business and the overall number of projects that you complete per year.

It is not necessary to debrief all projects. Who has the time and capacity to process all the information? Instead, carefully develop a set of criteria that will help you maintain consistency in your choices.

Examples of Criteria for Selecting which Projects to Debrief

- Projects that were strategically important to your organization.
- Projects that failed to meet the documented objectives.
- Projects that were exceedingly successful.
- Projects that were particularly complex.
- Projects that were not traditional or typical — i.e., different from the majority of projects you lead.

Customize this list and generate your own priorities. Collaborate with your team and managers to develop consensus around a few criteria that will allow you to select only high-leverage projects to debrief.

Postmortems

Critical Definitions and Concepts

Example: Selecting the Right Project to Debrief

The template below is an example of a simple tool you can use to select the right project. You can further develop the template to track, prioritize and optimize the results from debriefs.

Project Candidate	Business Case	Focus (Topic)	Plan Leader	Target Date	Deployment Leader
Selection of projects is based on a clear criteria, which is applied to all candidates.	The business rationale that justifies the expense in resources to perform the debrief.	The specific area of focus for the debrief. The desired outcome will dictate the focus.	The name of the person responsible for leading the planning and execution of the debrief.	The intended date when the main session of the debrief is planned.	The name of the person who is responsible to lead the implementation of the results of the debrief.
"Eagle Eye" platform	The investment in developing the new platform was over 10 million dollars. We must understand the effectiveness of the investment during the development cycle.	We will focus on our relationship with our vendors as 40% of the new platform was outsourced. We will also focus on the use of the new CAD software.	George Smith, VP of Supply Chain will lead the debrief process.	We will debrief two weeks after the product is announced to the market.	Mary Carter will be responsible for ensuring the deployment of the finding and continuous improvement.
"Columbia" (new product)	This project will not be debriefed as it is a simple improvement to an existing product.				
"NewWave" (new product)	This is our first product where more than 50% of the product is being outsourced to vendors across Asia. We must identify areas of improvement in our supply chain.	The focus will be in three areas: • Contract negotiation and execution. • Technical documentation. • Schedule management.	Lani Lui will lead the debrief; he is the project manager.	We will debrief twice. • At the mid-point of the project. • After the product has been rolled out to the market.	No person has been identified for deployment. For now our Quality Manager, Miguel Martinez, will monitor and make recommendations.

Product Development

Critical Definitions and Concepts

Second Critical Decision: Which Specific Improvements Will You Implement?

The second challenge is selecting the material from the debrief that you will actually deploy. Every debrief will yield a considerable amount of information, but only a fraction of it is worth deploying.

Here again, the development of selection criteria is important. An important step is prioritizing the information collected from your debriefs. As we outline in the process map for debriefs, every debrief should result in an action plan. In addition, every debrief will yield raw data — that is, information that is important but has yet to be ranked and sorted.

Processing all the information from your debriefs is a critical step. Processing means sorting by relevance to the organization, audience or constituency. For example, some information may be important to your marketing organization but not relevant to your finance group.

Examples of Prioritizing the Information

- Important and urgent. This is a two-variable criteria. The decision is made first by how important the improvement is to your organization and then by how urgent it is to deploy the specific improvement.
- Ability to leverage to improve productivity — i.e., which information will deliver the most performance improvement?
- By function — e.g., marketing, engineering, manufacturing, etc.
- Relevance to the executives of your organization. Allow the Executives of the company to prioritized based on the business objectives of the organization.
- Strategic information versus tactical.

Third Critical Decision: How Will You Effectively Deploy the Needed Changes to the Organization?

You must now select the method that you will use to deploy the information. What will be your strategy for ensuring that actual change occurs. In chaotic organizations it's always best to choose at least two ways to deploy. For example, through training and the quality manual. Or, through a web page and special presentations.

This is the most important step. It's the payoff for all the hard work. Consequently, you must dedicate considerable focus and effort. Failure to effectively deploy will render that debrief useless. Proper deployment will result in visible improvement in your organization's productivity.

Fourth Critical Decision: What is Your Strategy to Make the Needed Changes a Permanent Part of the Culture of Your Organization?

Finally, how will you ensure that change is long-lasting? So, you've decided how the valuable information will be deployed. But you are not done yet! The most often-forgotten step is tracking the deployment and providing enough support to ensure that the change A) does takes place and B) is long-lasting.

We suggest that you first form a Core Team to help you. This is a small group of people who volunteer to help you because they believe that the specific change you propose is important to the organizations. As a Core Team, hold a meeting to develop your action plan to sustain the change. Remember that at this point the change has been deployed (see the previous Critical Question).
Something else that will help sustain a change is to deploy a very simple metric to track progress. You can also request support from your manager and from the Human Resources personnel.

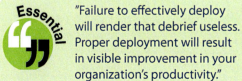

"Failure to effectively deploy will render that debrief useless. Proper deployment will result in visible improvement in your organization's productivity."

For comments, questions or additional copies please e-mail the author: debriefs@rapidinnovation.com.

Conducting Debriefs

Key Questions to Consider

Will We Need a Facilitator for Debriefs?

In a Word, Yes

A facilitator is a neutral person who has the skills to lead your team in a productive search. A facilitator can create a safe environment, reassuring your team members that the intent is to improve performance, not to punish anyone. A facilitator focuses on the debriefing process, ensuring that your team is productive. Finally, a facilitator has the skills to expedite the process and focus your team on the right thing at the right time.

You can hire an outside facilitator. Another alternative is to ask your HR department to help. Or you could invite another project manager to facilitate your meeting. Perhaps a manager from a different organization might fit the bill. Do your best to secure a facilitator because it makes a considerable difference.

What About Multi-Site Projects

Product development for many companies takes place in many locations around the world, and this practice is growing. When separated by distance, language and culture the need for debriefs becomes more important. We strongly encourage you to do more frequent debriefs if your development team is scattered around the world.

The purpose of a multi-site debrief is the same as with a co-located one: Improve the future performance of your organization. The preparations for a multi-site debrief vary based on the distance, language and culture. To reduce travel costs, add debriefs to the last project meeting, which is already budgeted. Furthermore, there may be reasons that warrant having the team travel for a debrief — for example, a project that was exceptionally successful, one that failed or a project that represented a significant new strategy for your company.

Using Video Conference to Do a Debrief

If you cannot meet face-to-face, then you must prepare the logistics carefully. The next best thing to a face-to-face meeting is a video conference, as it allows for some interaction. The process of selecting which project to debrief and what part of the project to debrief is the same as with any other. The same can be said for what you will do after the debrief. Here are some steps to take initiating the process:

1. Start by selecting a volunteer in each of the locations who will serve as the "point person"; that is, your representative in that site. He or she will help with logistics, preparations and follow up after the meeting.

2. Schedule the meeting and the video facilities.

3. Send clear notices and talk with each team member — preferably by phone rather than e-mail — so they understand when and where to meet.

4. Prepare the team by providing the right information prior to the debrief — for example, the objectives of the debrief.

5. Ask each team member to bring to the video conference a list of two things that worked particularly well during the project and two things that must be improved.

6. Use the same process to do the debrief taking into account the use of video. You want to create as much interaction as possible.

7. You need to have someone capture the information electronically, as you may not have the luxury of yellow stickies. You can use a spreadsheet to capture and prioritize the feedback. Prioritization may be done after the video conference to save time and money.

Be mindful of cultural and language differences. Some cultures are not comfortable with anything that appears to be negative criticism. Coach team members that the debrief is about the project, and not about any individual. Encourage them to write down the information prior to the meeting to remove some of the stress.

Language represents another challenge. Many people for whom English is a second language tend to read better than speak. This is another situation where asking them to write their input prior to the video conference is helpful.

You may want someone local to collect the information and present it during the meeting. This way other team members don't feel put on the spot.

The Role and Responsibilities of Managers and Executives

1. Managers and executives should avoid attending the debrief meetings. Allow the owners of the problems to solve them. Also, too many leaders may result in stifling the flow of ideas. Finally, managers and executives may tend to intimidate due to fear of punishment.

2. Managers and executives must be ready to provide support, funding and encouragement once the solutions have been identified by the team in the action plan.

3. Managers must avoid any blaming or guilt during any stage of a debrief, which would inevitably have a negative impact during future debriefs to the point of rendering them useless. All debriefs are guilt-free! Use other means to appraise performance.

4. Celebrate the genuine improvement in productivity due to debriefs. Remember that every dollar saved though productivity goes directly to the bottom line!

Getting to the Root of Things

During your debriefs, you will receive a set of symptoms — for example, "The document tracking software did not work." Naturally the question begs, "Why did the document tracking software not work?"

You need to be familiar with methods of root cause analysis such as the Fishbone Diagram or the Affinity Diagram. This way you can help your team identify the root of the problem, which will enable you to fix things once and for all. Additionally, The 5 Whys is a simple way to facilitate identifying the root cause of problems.

The moral of this story is to beware not to fix symptoms, but instead fix the cause of the problem.

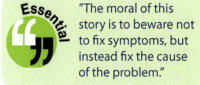

The Power of the Lowly Yellow Stickies

These little pieces of paper are tremendously helpful in making your debrief more manageable and productive. Ask your team members to record their input on yellow stickies; for example, Post-It Notes® (3M brand, model #655 3x5). This creates a documented record that can later be used to organize and prioritize the information.

Your facilitator will be familiar with several ways to use the yellow stickies to make decisions, organize and prioritize all the information from your debrief. While we're on pieces of paper, the use of flip charts (easels with large pads of paper) is also very important for debriefs. Make certain that you have an ample supply of these valuable aids.

The Protocol to Describe Debrief Input

It saves considerable time and improves the overall quality of your debrief to establish and enforce a protocol to describe the input from your team. When you start your meeting, describe the protocol, show a few examples and tell your team to start using it. During the meeting you will probably have to remind them to continue to use the protocol. You need descriptions that are clear. This debrief may be your best opportunity to obtain this input, and you might not get another opportunity.

Conducting Debriefs

Examples: Effectively Describe Issues and Opportunities

The Protocol

- Description of the issue or opportunity.
- Description of the possible root cause.
- Description of the impact, damage or improvement.

> Describes what went wrong or what went right. (No names of individuals!)
>
> Describes the root cause of the problem. If you don't know it, then leave this section blank.
>
> Describes the impact, damage or opportunity. Remember, debriefs are also about things that went well and should be repeated.
>
> Author's Name

Example 1

- Description of the issue or opportunity.
- Description of the possible root cause.
- Description of the impact, damage or improvement.

> The Operators Manuals were not delivered on time due to a delay at the printing plant in India, which forced us to ship incomplete and poorly printed Operators Manuals for the first 1500 units.
>
> M. Change

Example 2

- Description of the issue or opportunity.
- Description of the possible root cause.
- Description of the impact, damage or improvement.

> The first prototypes were delivered on time, were faithful to the design and had excellent detail. We used a new supplier to try a new Rapid Prototyping machine. The speed and quality saved us at least two weeks in the schedule and provided a higher quality prototype.
>
> L. Medina

Product Development

Conducting Debriefs

Rules and Strategies for Debriefs

Strategic Rules

1. The key question to ask before you do any debrief: What is its intent? You need a specific, documented purpose that shows why this debrief is important, why this focus for this project is worth the effort.
2. Before any debrief, ensure that there is a willingness to adapt and improve in your organization. Every debrief should result in some type of change.
3. All debriefs should reach specific, documented findings and culminate with an action plan or roadmap for improvement.
4. A debrief should be about how teams and organizations worked, not how individuals worked.
5. Debriefs are always guilt free — they are not performance reviews, and no one should be punished or reprimanded.
6. Not all projects need to be debriefed. Carefully select those that might yield considerable leverage.
7. All debriefs are done to create the capacity to improve future performance.

Tactical Rules

1. Debriefs should be planned and carefully choreographed to ensure quality input — they are not casual events.
2. The best tool for debriefs is yellow stickies (Post-It Notes®).
3. Debriefs should allow ample time for dialog and discussion — generally two to three hours for interim debriefs and a full day for end-of-project.
4. A debrief should always be a positive experience. Without open communication and common intentions, the effectiveness of the debrief will be diminished.
5. Debriefs can be held throughout the project, not just at the end.
6. Debriefs should include all team members whenever possible. It is the perception of the entire team that provides the insightfulness you desire.

Five Key Questions Before You Conduct Your First Debrief

1. Are the Executives of your organization in support of debriefs?
2. Do your team members have the training needed for successful debriefs?
3. Do you have a long-term plan to sustain debriefs for the long run?
4. Do you have the tenacity to stay with it?
5. Do you have a plan to demonstrate the value of debriefs once you start?
6. Do you know what you will do with the information from a debrief in order to improve productivity?

See Also: Deployment

Conducting Debriefs

Postmortems

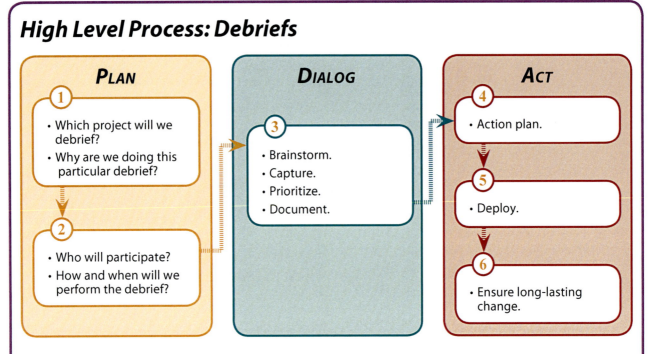

The above diagram shows a high-level process for debriefs. Note that it all starts with identifying the reason for your debrief. It is also important to know if you will have the means to implement the changes. Your team members and the rest of the organization will be disappointed if they do a debrief only to find out that none of you have the time, resources or support to implement the changes.

Detailed Process: Debriefs

WHERE YOU'RE AT

At this point you have used the criteria for selecting a project to debrief. Not all projects need to be debriefed — you must have a clear reason for selection.

The assumption is made that the project you have selected meets the criteria.

1

SET OBJECTIVES

Decide what you're going to debrief — for example, the process, the communication, leadership, flow of documentation or change notices. Or do a general debrief, open to all issues.

Decide what you will do with the information once you have it — i.e., how will you ensure that the input translates into action and eventual improvement.
DO NOT DO A DEBRIEF WITHOUT THIS STEP!

See Also: **Debrief Planning Tool**

2

DOCUMENT YOUR PLAN

Use the Debrief Planning tool to help you prepare.

CONTINUED

Product Development

Conducting Debriefs

CONT'D

Ensure that you invite the participants. Inviting is much more than sending an e-mail.

You need to talk to each person, and ensure that he or she has your debrief on the calendar. Reassure everyone that their input is valuable.

You might want to ask the managers, when they hold their staff meetings, to remind people to attend your debrief.

SCHEDULE THE DEBRIEF MEETING

A good debrief should last at least three hours to allow for proper discussion, prioritization and action planning. For large and complex programs, a debrief can take as long as a full day, and in some cases longer.

Decide whom you need to invite given the objectives of the debrief. It's better to be inclusive rather than exclusive.

Collect relevant documents that can be used to help in the debriefing process. Make copies for everyone as appropriate.

Create an agenda — for example:
- Set the stage.
- Tell participants what will be done with the information.
- Review the history of the program to refresh people's memories.
- Open discussion using yellow stickies.
- Identify actionable items — i.e., items worth taking action on.
- Prioritize.
- Assign responsibilities for action.
- Agree on the next step, to ensure closure.

Prior to the start of your debrief you developed an agenda. But sometimes even with an agenda it's difficult to get things started.

You might want to rebuild the path of your project — i.e., use a chronological approach.

You might want to use the functional approach — i.e., talk about the marketing, engineering or financial aspects of your project.

Another possibility is to start with the positive parts of your project — e.g., what worked?

You have the option of asking every participant to come prepared with two things that went particularly well and two other things that need improvement. This will help jump-start the conversation at your meeting.

Select a large conference room to allow team members to break into small groups as needed.

Ensure that you have all the office supplies you will need; i.e., yellow stickies, black felt tip markers, flip chart easel and paper.

CONTINUED

For comments, questions or additional copies please e-mail the author: debriefs@rapidinnovation.com.

Conducting Debriefs

Postmortems

WHERE YOU'RE AT
At this point you have begun the debrief meeting.

CONT'D

④

SET THE TONE FOR A POSITIVE EVENT

Reassure the participants that no one will be punished, reprimanded or otherwise sanctioned as a result of this debrief. A debrief is not a performance review.

It is important that the entire team understand and agree that this meeting is not for punishing, but creating new opportunities, improving performance and reducing future stress — in short, it is a positive event.

Create a safe environment where people feel they can speak freely without fear of retribution.

Use yellow stickies to get the information out. Many people are not too eager to speak in a larger group.

PEOPLE MUST FEEL THAT THEY CAN TRUST YOU AND THE PROCESS!

Avoid the blame game. Stay away from "right" or "wrong;" rather, focus on "worked" or "didn't work." Also, start by listing achievements. This will reduce stress.

Don't use people's names; rather, talk about the problems, issues and breakdowns.

Ask "What should we do differently next time?" This will create a forward-looking environment and reduce the tendency to look back in blame.

⑤

REVIEW THE PROJECT

Start with a quick review of the project. This is an opportunity to share the materials you've prepared; for example, the schedule and other documentation.

This is also a good time to remind your team of the specific focus; for example, communication, supply chain or technical development.

Create a Parking Lot

This is a handy tool to capture information that is important, yet not relevant to the focus of your debrief.

Take a sheet of flip chart paper, label it "Parking Lot," and adhere it to the wall.

Encourage your team to add yellow stickies as the meeting progresses.

The Parking Lot will reduce the number of distractions. Plus, it will help you capture valuable information for later use.

⑥

FACILITATE BRAINSTORMING AND DIALOG ABOUT WAYS TO IMPROVE

Use yellow stickies to document all input.
Some questions that might create a lively discussion:
- What worked?
- What didn't work?
- What should never be repeated?
- What should be emulated in future programs?
- What tools worked well? What tools did not work?
- What training was useful?
- What additional training is needed for future programs?
- How effective was the communication loop?
- What abilities or skills need to be improved?
- What created the most distraction? And what will we do differently next time?
- What one thing slowed down the program the most? And what will we do differently next time?

CONTINUED ▸

For comments, questions or additional copies please e-mail the author: debriefs@rapidinnovation.com

Product Development

Conducting Debriefs

WHERE YOU'RE AT

At this point several hours may have passed.

You will have collected a considerable number of yellow stickies. All of this is raw data as — it is not prioritized nor organized in any fashion.

It is clear that no more valuable input can be collected.

CONT'D

(7)

ORGANIZE AND PRIORITIZES NECESSARY IMPROVEMENTS

Once you have covered your agenda, it's time to organize and then prioritize the input using consensus.

You must perform this step before declaring the session over. It is very important!

There are several ways to approach this critical step. For example, use the affinity diagram or story-boarding approaches. You might want to bring in a facilitator who is experienced in the use of these tools.

Prioritize all the input using a clear and simple set of criteria. For example, urgent and important, most needed for the next program or highest overall leverage.

See Also — **Critical Tools**

(8)

DEVELOP MESSAGE TO MANAGEMENT

At the end of the discussion (brainstorming), ask the team to develop a message to management: What one thing does the team wish management to know about this program? For example, perhaps it is something that needs to be improved but is beyond the responsibility of your team.

This is a separate document or presentation you and your team develop for your managers and executives. It is important because it engages them in the debriefing process. Plus, it's your opportunity to ask for support and resources to implement needed changes.

Your message should be clear but not confrontational. Avoid blaming management; rather, present your information as an opportunity and in a positive vein.

This message can be delivered to your management by a small group from your team, or you can do it yourself. Another possibility is to ask a willing manager to deliver the message during an appropriate staff meeting.

This is an optional step but one we highly recommend. Your executives will welcome your suggestion. Moreover, management can be very helpful in ensuring change that is long-lasting.

 Important — Your Message to Management needs to include your triumphs and successes! Debriefs aren't just about what needs to be improved but also all the good things that took place. Ensure that your managers and executives know about your team's results and breakthroughs.

And it's worth repeating: Keep all your reports and presentations blame free! Performance issues and incompetence can be dealt in separate forums — your HR people can help.

IMPORTANT NOTE: "Incompetence" simply means someone didn't have the skills, knowledge or ability to perform a job. It doesn't mean he or she is a bad person. We recognize that the term "incompetence" in most contexts has a negative connotation. We are using the term as neutral, and we expect you to use it the same way.

We cannot omit the statement, but we can rephrase it: "lack of competence," "lack of fitness" or "lack of abilities."

 Important — This is a good opportunity to also review your Parking Lot and extract relevant information in your Message to Management.

CONTINUED

29

Postmortems

Conducting Debriefs

CONT'D

9

CONCLUDE THE DEBRIEF MEETING

At the end of your debrief meeting you will have:

- A documented and prioritized list of improvements, be it "do more of" or "do less of."
- A list of ad hoc, tactical improvements with clear delivery dates and responsibilities. These are simple improvements that do not require much effort but can deliver much value.
- A Message to Management document.
- An archive of important information from your meeting, beyond the list of improvements.

See Also: **A Well Documented Action Plan**

10

CREATE AN ACTION PLAN

During the debrief, you organized and then prioritized the list of suggested improvements to be made. Now you need to translate this list into a roadmap — i.e., the action steps needed over a timeline.

As with any action plan, it must document clearly what needs to be improved, who is responsible for ensuring the improvement and when the improvement will take place. Otherwise, there is a high probability that all the effort will be wasted.

Key points:

- It may require a subsequent meeting of key players.
- Make certain that the plan is developed in collaboration with your team members.
- Distribute segments to the appropriate stakeholders — i.e., pass out the action items from the plan, but do not flood team members with information that is not relevant to them.
- Decide and document how you will know the changes have taken place.
- Don't forget the critical questions:
    – What must we do differently next time?
    – How will we make certain we change for next time?
- Ensure that all relevant information is documented and archived for your future use and for the use of other projects.
- You must create the closed loop to ensure the benefits of your efforts. Once all the action items are completed, you want to ensure that the good work done results in actual change — improvements — in the organization.

CONTINUED

For comments, questions or additional copies please e-mail the author: debriefs@rapidinnovation.com.

Product Development

Conducting Debriefs

WHERE YOU'RE AT
At this point you have concluded your meeting and everyone has gone back to their jobs. It is now time for you to lead your action plan to results!

CONT'D

⑪ IMPLEMENT THE ACTION PLAN

This is a case where dogged tenacity wins the day. You must stay on top of the action plan in order to ensure completion and, most of all, productivity improvement.

This is the roadmap that you and your team create to change things that are within your control. Generally, this is known as remedial action.

You will find areas of improvement in processes and systems beyond the control of your team. Ensure that your action plan differentiates between the things that your team will improve and those that need the attention of others.

⑫ DEPLOYMENT

This is not the same as the action plan. It is the process by which you let every one else in your organization, your company even, know what needs to be improved.

Also, this is the place where you communicate good practices so other organizations benefit from your findings.

This information must be communicated proactively. You might want to recruit others who participated in the debrief to help you accelerate change.

This will challenge your commitment to make a difference. It is up to you to track the progress, and lead the organization into taking action.

⑬ KNOWLEDGE BANK

Critical information from all of your debriefs should go into a single storage area, which is called the Knowledge Bank or Best Practices.

This makes it convenient for your entire organization to access the results from all of your debriefs.

This is not a trivial function, as the data base must be kept up to date, prioritized, sorted and, most of all, convenient to your organization. It must not be some obscure folder on some remote server.

⑭ LONG-TERM IMPROVEMENT

Once you have developed your action plan and you have done the job of disseminating the information, you are still not finished.

Many of the improvements and good practices that you discover require a cultural change. That is, you must lead the organization to drop old habits and adopt new ones.

You need to work in concert with your managers, executives and HR department to ensure that change does takes place.

LEARN AND IMPROVE (KAIZEN)
Learn and continue to improve your process. Over time you and your team will develop the needed expertise to perform debriefs that make measurable difference in your entire company.

For comments, questions or additional copies please e-mail the author: debriefs@rapidinnovation.com.

Conducting Debriefs

Create a Well-Documented Action Plan from a Debrief

Warning

The biggest challenge about debriefs is carrying out the action needed to improve productivity. It is relatively simple to schedule the debrief with your team; it is difficult to ensure that actual change happens as a result of the debrief.

Before you call a debrief, ensure that you have the time, resources and support to implement the changes that will be expected after the debrief itself.

To ensure change you first need to develop an action plan — a detailed document with clear descriptions of the action items, unambiguous appointments of who will be responsible for carrying out the action items and a timeline indicating when the action item will be completed.

Example: A Well-Documented Action Plan

Item	Background Info	Action Item	Responsibility	Timeline
1.	Key specifications were not documented, which caused three weeks delay in the "Eagle" project. *(The required action is clearly stated to avoid confusion.)*	The product marketing managers will call a review of the specifications of the MRD (Marketing Requirements Document), starting with the next project and before the MRD is sent to engineering.	Lou Chin, Product Marketing Manger. *(Responsibility is clear by writing the name and title of the individual.)*	Lou will report progress to the Core Team no later than October 10, 20XX. Lou will complete the action item no later than October 21, 20XX. *(The action plan is not complete until clear dates are included.)*
2.	There was confusion about the schedule during the "Eagle" project. *(This column explains the background of the problem as it was discussed during the debrief. This is helpful for reference.)*	Create a large schedule Gantt chart using a white board and yellow stickies so it is visible to everyone on the team.	Mary Lynch, Engineering Manager.	Mary will work with the engineering team to create the schedule by the start of the next project (Eagle 2, January 2, 20XX).

The above example shows a well-documented action plan from a debrief. Do not call a debrief meeting if you do not have the resources to ensure the completion of the action items. A debrief has value only when it results in visible improvement in the productivity of your organization.

Product Development

Conducting Debriefs

Example: Debrief Planning Tool

To be filled out by the person responsible for organizing the upcoming debrief.

Item	Description
Project to Debrief	Project "Eagle" will be debriefed because the supply chain model failed and caused a six-month delay.
Brief Description	"Eagle" was a project to develop a new product platform to address the needs of the sheet metal industry, which represents a new market for our company.
Specific Focus	We will analyze the schedule and the supply chain to extract and prioritize improvements for future projects. We will also analyze our Quality manual and the supply chain management function.
Date, Time and Location	November 5, 2009 Conference room #210 We will start promptly at 9:00 a.m. and finish at 12:00 noon
Attendees	• Project manager • Product manager • Engineering manager • Manufacturing line manager • Supply chain manager • Base data manager • Logistics manger
Reference Materials and Documentation Needed	• The project schedule • A list of the major delays and their impact on the overall schedule • The list of suppliers in the supply chain • A "block diagram" of the "Eagle" product • A unit of "Eagle" disassembled to allow the examination of key components

Deployment and Long-Lasting Changes

How to Actualize Productivity Improvement

The Biggest Challenge of Them All

If there is a common problem with most if not all debriefs, it is the way the information is translated into productivity improvement (deployment and adoption). The intent of all involved in debriefs is to cause change, so they and their peers don't have to suffer through the same problems. Yet most debriefs result in little or no change.

This is not the fault of the people doing the debriefs. It is the lack of a clear process, and sometimes the lack of leadership, that causes the dead end. Consider that there are really two major stages to every debrief. One is obtaining and prioritizing the information. The second is the transfer of that information in a way that causes change.

We suggest you develop the process to deploy the information before any debrief takes place. This can be done by a small team, a group of project managers, a facilitator or your HR department.

Key Questions when Developing the Framework for Deployment

- Who will collect and archive all the information from all the debriefs?
- How will that information be categorized for easy retrieval and future use?
- How will the decision be made as to which improvements will be deployed throughout the organization?
- Who will make the above decision?
- What will be the process to deploy the information?
- What will be measured to ensure that the deployment is working?
- What will be measured to ensure that improvement is taking place as a result of the debriefs?

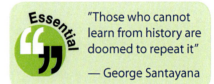

"Those who cannot learn from history are doomed to repeat it"
— George Santayana

Who Can Help?

- Recruit your HR people to help develop an action plan to ensure progress.
- Create a team (Quality Team, Continuous Improvement Team, etc.), and charter it to carry out the change over a set period of time. These are small teams of three to five people who can implement the specific changes. These teams are generally made up members from your core team, but you can tap into other groups when it makes sense.
- Get the executives of your organization to commit to carrying our your recommendations. They can turn them into an "initiative" as part of their business or operational planning process.
- Some organizations have a quality manager who is responsible for ensuring continuous improvement.
- Form your own "Tiger Team" among the people who participated in your debrief, and help them carry out the needed changes.
- Make a commitment to drive some of the change yourself. You'd be surprised how much one individual can influence an entire organization.
- Work with your training organization to incorporate the changes into the training of new and current employees. Your training department can also incorporate changes in the documentation used for development.

Product Development

Deployment and Long-Lasting Changes

Spreading the Knowledge

Once you've prioritized the information gleaned from the debriefs, it's essential to promote awareness and acceptance of the lessons learned. Don't assume your organization will change automatically or even quickly!

Below is a short list of how you can spread the knowledge. Based on your organization's culture and reporting structure, you and your team can find even more innovative ways to deploy information so it creates long-lasting change. The key question remains: What is the best way to ensure performance improvement?

Examples	Description
1. Training and Development	Continually update your training materials to reflect the improvements. Ensure that all new project managers are trained on the lessons learned, and run incumbent project managers through refresher training.
2. Brown Bag Sessions	Create a series of 60- or 90-minute sessions during lunch, breakfast or before or after work, facilitated by a project manager. Offer a meal or refreshments.
3. Good Practices Web Page	Develop a "good practices" (some companies call them "best practices") web page for your intranet, extranet or internet site. Make full use of the internet to keep everyone fully informed and trained, including those who are in other states or countries.
4. Community of Practice or User's Group	Start a group of people who volunteer to meet regularly on a particular topic because they have a passion for said topic. The organization provides support in the way of facilitation and resources.
5. Executive Retreats	Planning meetings, retreats and off-sites can include time to present the lessons learned. But you must be proactive and have much tenacity to prevail. Get your Management Mentor to schedule the presentation. Better yet, show them your leadership by scheduling it yourself.
6. Staff Meetings	Schedule a presentation for 15 minutes at the start of your staff meetings, and give it priority. Or dedicate one full staff meeting each quarter to discuss your debriefs. Be aggressive, and schedule a half-day staff meeting just on debriefs.
7. New Projects	At the start of all new projects, allocate 60 to 90 minutes to talk about lessons learned from previous projects or review good practices.
8. Required Reading	Document improvements, make them convenient for all to read and then make reading them a requirement — no excuses!
9. Tools, Templates and Guidelines	Incorporate performance improvements into tools, templates and guidelines. Constantly update your documentation to reflect the lessons learned.
10. Quality Manual	Incorporate performance improvements into your quality manual. Use Six Sigma, ISO-9000 or your quality system to ensure that changes are incorporated in the processes and SOPs.
11. Special Meeting	Have a special meeting after the results of a debrief have been documented. Dedicate quality time to discussing the recent findings, and invite managers. This is done before any documentation is published — do it ad hoc or extemporaneously.
12. Project Management Excellence Meeting	Once or twice a year hold a Project Management Excellence meeting. This can be a half-day or even a full day to share the improvement and engage the entire organization in continuous improvement.

Deployment and Long-Lasting Changes

Example: Debrief Planning Tool by Focus (Topic)

This is a comprehensive form to enable conversations. It may be too complex for some projects or programs, so we suggest that you customize this form.

Do not send this form to team members! This form is to help you facilitate the meeting, but the real value is the dialog.

1. Planning

   For example:
   - Were the goals, purpose and objectives clear to the team?
   - Was the plan for the project sufficiently clear to enable the team to deliver results?
   - Were the executives and managers available to clarify the goals of the project?
   - What steps will improve the planning of future projects?

2. Resources

   For example:
   - Were the resources properly planned?
   - Were the contractors, temps and consultants available when they were needed?

3. Scheduling

   For example:
   - Was the initial schedule realistic?
   - Were the project milestones helpful in creating progress?
   - What were the obstacles to meeting the schedule?

4. Performance of Project or Program Management

   For example:
   - How did the project plan work (WBS)?
   - How well was the project defined?
   - How well was it tracked?
   - How well was the plan communicated?
   - How well was the plan documented?
   - How widespread was the use of the project plan?
   - Was the plan realistic from the start?

Product Development

Deployment and Long-Lasting Changes

5. R&D / Technology Management
   For example:
   - How was the project managed at the R&D level?
   - How well did R&D communicate methods, technology, barriers and risks?
   - Any breakdowns in R&D? Why?
   - Were technical issues identified early enough?
   - Were technical issues communicated early enough?
   - Was there an effective method to address technical issues?
   - Were estimates of time and cost on track?
   - What interfered with R&D?
   - What helped R&D?
   - What were the lessons learned?

6. Communications
   For example:
   - Were all stakeholders informed?
   - Did the team communicate effectively (the right information to the right team member at the right time)?
   - Were remote team members properly integrated in the communication loop?
   - How about team members in other countries (language, culture, timing, etc.)?
   - Was the communication with managers and executives effective?
   - What must be done next time to improve communication?
   - What should we never do again when it comes to communication?

7. Managers and Executives
   For example:
   - Did your managers help the team be successful?
   - Were decisions made on time?
   - Were decisions properly communicated to the team?
   - What should managers and executives do next time to improve overall performance?

8. Product
   For example:
   - Was the product delivered to the customers on time? Was it available to order on time?
   - Was the introduction to the market place effective? As a public announcement? Through sales channels training? Through major accounts? Internationally?
   - Was the promotional material available at the right time?
   - Were the customers satisfied with the product rollout?

Deployment and Long-Lasting Changes

9. Marketing
   For example:
   - Was the product launch effective? Says who?
   - Was the product proposal effective? Says who?
   - Did the market message work? Says who?
   - What slowed down the introduction?
   - How was marketing handled in the venue of: Sales training? Promotional materials? Internet pages? Major accounts? International?

10. Manufacturing (In Software Development, use Documentation or QA)
    For example:
    - Was the team properly prepared to receive the product?
    - Did the product meet the manufacturability targets?
    - What was the root cause of any unexpected delay?
    - Was the BOM (Bill of Materials) accurate?
    - How about the supply chain? Did it work as it was planned?

11. Virtual or Distributed
    For example:
    - Were the tools used to work with distant team members appropriate?
    - How about document management? Engineering documentation?
    - Was e-mail valuable? Why? If not, why not?
    - How about project management in distant locations?

12. Service and Repair
    For example:
    - Was there an effective service plan?
    - Was it on time?
    - Was it communicated clearly?
    - How about service training, service options, international service, spare parts, repair manuals, diagnosis, metrology and bud tracking?

13. General Questions
    For example:
    - What went exceptionally well?
    - What went exceptionally poorly?
    - What was the most important lesson we learned?
    - What should we never do again?
    - What should other projects remember about ours?

Critical Tools

PRODUCT DEVELOPMENT

Deployment of a New Tool 40
Deploying a tool takes extra work. You are not just enabling the adoption of a new tool or method, but actually changing old habits. Don't be daunted by setbacks; you're doing the right thing for your team and company.

Yellow Sticky Protocol 47
This is a way to track and code information to enable future processing. It's very useful when you have a large number of stickies or many people contributing.

Affinity Diagrams 53
We are not linear thinkers. We tend to jump around when brainstorming. Using Affinity Diagrams is a simple method for organizing the brainstorming session — the non-linear flow of ideas — into useful, documented facts. And it is a fine way to ensure consensus.

Prioritizing Affinity Diagrams 57
Prioritizing is all about making decisions and reaching consensus. The secrets are selecting the right criteria to use in making those decisions and clear communication.

Problem-Solving Process 60
Don't underestimate the need to apply a formal process to solving a problem — simple or tough!

Karnaugh Maps: 2x2 Matrixes 65
Help your team make a decision or a selection to set priorities from an array of many options; for example, determine the risk factors of a new deployment or select key features on a new product.

The 5 Whys 72
This is a simple methodology to obtain valuable information from customers in order to get to the root message.

The Fishbone Diagram 75
It's tempting to try to solve the symptom of the problem without trying to analyze the real cause. This is a fine tool for providing the discipline to go from "symptom" to actual "root cause" of the problem.

Process Mapping 82
A process is a documented, sequential list of value-added steps, the aggregate of which is used to deliver a product to a customer.

Deployment of a New Tool

Cascading a New Tool to the Organization

Introduction

There is a difference between attending a seminar or workshop and causing your organization to embrace any or all of the tools you learned. Said another way, just because you and a number of your peers attend a seminar, that's no guarantee that what you learned will become common practice in your organization.

Deploying a tool takes additional work. It is a process, and requires planning, tenacity and leadership. Consider that you are not just enabling the adoption of a new tool or method, but actually changing old habits. In short, you will cause disruption and a break from the comfort zone currently in existence in your organization.

The Business Case

The return on investment (ROI) from any seminar or workshop only occurs after the class and when visible and measurable improvement in productivity is achieved. Some have the erroneous notion that attending a seminar is the objective; instead, attending is a means to and end. The real objective is to take what was learned and apply it to improve productivity.

Expect Resistance to Change

However, just because you and a number of your peers attend a seminar or workshop, that's no guarantee that what you learned will become common practice in your organization. Deploying a tool takes additional work. It's a process that requires planning, tenacity and leadership. Consider that you are not just enabling the adoption of a new tool or method, but actually changing old habits. In short, you will cause disruption and a break from the comfort zone currently in existence in your organization.

Expect resistance to change in overt and passive ways when you suggest the adoption of a new tool or method. You will hear things like "It won't work here," "We tried that, and it didn't work," "Our situation is different," and many similar statements. There are numerous resources about driving change in an organization to help you.

The Hesitaters

The initial push-back may actually be the expression of other feelings:

- **SEEKING ADDITIONAL INFORMATION:** Some in your team resist as a way to obtain additional information, as they feel fearful to deploy something they do not understand. It's the best way they know for doing fact-finding.
- **SEEKING REASSURANCE:** They feel unsafe; for example, they want to know that they will not get fired if the tool does not work the first time out.
- **TESTING:** They are simply trying to determine if this is for real and if you have a strong conviction that the tool will work.
- **GENUINE CONCERN:** Some may actually feel that the tool will not work or is not applicable.

The way to deal with all the above feelings is to first be aware that they are common, and then validate the actual source of the concern by providing additional information, reassurance, etc.

The Reactionaries

Some individuals are simply opposed to change, reactionaries who want to preserve the status quo. Occasionally you will run into someone who will refuse to change, refuse your new tool and steadfastly remain opposed. It is rare that you are able to change these individuals, and we don't have the time here to

Much that has been written about driving change in an organization. When searching for books, seminars, articles and other resources use the terms "Managing Change" and "Resistance to Change."

Also, your human resources department may be able to help or offer additional resources.

Product Development

Deployment of a New Tool

delve deeply into this topic. We suggest that you do not become discouraged; rather, build a constituency of those who do want to change, and use it to overcome the resistance.

Food for Thought

The reluctance to change is a part of the human condition; you are not a bad person if you resist change. Your strategy to overcome resistance must take into account that it is a normal reaction. Further, the process to overcome resistance is all about people. Do not rely on templates, software programs, process maps, etc. Rather, seek to understand the situation, and then tap your leadership skills to drive the change.

Here are a few things to keep in mind:

- First, set about understanding the current situation of those who need to adopt the new tool or method. Talk to them as a group and, more importantly, individually. Make an effort to listen with the intent of understanding.

- Be sure to consider WIFM (pronounced "wiffem"), which stands for "What's in it for me?" We react positively when we understand our personal benefits. You must make the case for embracing the change such that every person on the team sees his or her benefits as part of the deal.

- Stop selling and start communicating! Our excitement for the new tool gets the best of us, and we start pushing the idea to the point where it becomes annoying. Instead, remain open to suggestions. Some will try to help you succeed by providing ideas for improvement — for example, a better way to use a template or a better way to deploy the tool. Be open to ideas, and engage those who provide them; do not reject them either directly or passively.

- Speak their language. Adapt your message to the audience. Speak in a way that is familiar to them. This will improve communication and avoid the use of new words.

- Start small. The intent of the first use of your new tool or method is to train and to develop the organization; thus, do not select a major project. Rather, find a simple project where the use of your new tool will shine and the risk of failure is very small.

- Finally, your tenacity and leadership will carry the day. Don't be disappointed with setbacks. In the end you're doing the right thing for your team and your company.

The Critical Questions

All of us need a certain amount of information before we commit to a tool, method or practice. It is not fair to expect people to jump at the opportunity if they don't have sufficient context, especially at the personal level.

Generally, we all want the following questions answered:

- Why are we doing this?

    (What is the relevance to my team, our company or our organization?)

- What would happen to us if we don't do it?

    (What would be the impact on or consequences to the company, organization or team?)

- Why are you asking me to do this?

    (Tell me why I should care. What's in for me?)

- What support can I expect from you?

    (Can I rely on you? What is your commitment to me?)

- What specifically do you expect me to do?

    (Tell me specifically how this impacts my job. What specific changes must I make?)

The above information must be communicated in the language and context of the listener. Avoid platitudes and irrelevant information as they tend to confuse and add uncertainty.

Deployment of a New Tool

Detailed Process: Deploying a New Tool

1

CHOOSE THE TOOL

It's best to start small. Deploy a single tool to reduce the challenge and risk of failure.

Define clear criteria; for example:
- Ease of learning.
- Amount of leverage the tool will provide.

You can select any criteria, but you need to be consistent in your evaluation and prioritization.

Start recruiting supporters and allies who react positively to the new tool.

 Use the Planning Document in this section to ensure that you are well prepared for the deployment.

2

IDENTIFY YOUR CORE TEAM

You will need a small group of your peers to help you deploy the tool.

Team members should be willing and able to spend the needed amount of time helping you with the task. Sometimes it may take up to four hours per week.

Make sure the team will be in place for at least three calendar months. It might take longer.

 Remember to secure the visible support of your organization's executives. Without it, you risk failure.

Meet with your executives, and show them the benefits and improvements that your new tool will bring to the organization.

If possible, secure a sponsor or a mentor — one of the executives who is willing to assist you in the deployment.

Seek support from others who are willing to help deploy.

3

SECURE SUPPORT FROM YOUR EXECUTIVES

It is prudent to identify one of the executives of your organization to sponsor your efforts.

Sponsors, also called Mentors, will provide moral support, help remove obstacles and add a measure of stature to your efforts.

The best sponsor is one who, like you, has a passion for the tool you are leading.

 As you select the pilot for your deployment, also ensure that you have training, templates and other needed support to help the team succeed.

CONTINUED

Product Development

Deployment of a New Tool

CONT'D

4. IDENTIFY THE PILOT

This is the team, organization or group of employees that will apply the tool to a real project.

The results of this pilot will provide valuable information when deploying the tool to the entire organization.

Remember to obtain approval from the appropriate managers and supervisors.

 Equip the Pilot

It is critical that you provide coaching, encouragement and support to the pilot team. There is the risk of push-back and disappointment if the tool does not work as expected

It would be best if you were part of the pilot team. That way, you will be there throughout the process.

5. CONDUCT THE PILOT

Ensure that you provide a high level of support to the organization such that the deployment of the new tool does not end in failure.

Provide all the training, tools, encouragement and other forms of support throughout the deployment of your pilot.

 What if the Pilot Fails?

Even the best of efforts may results in failure.

The first thing to do is to debrief and identify the root cause of the failure — without this step you cannot recover.

Avoid blaming, guilt or punishment as these things add no value to the recovery process.

Once you have the root cause and lessons learned, modify the process and try a new pilot.

6. DEBRIEF THE PILOT

The topic of the debrief is to identify what worked and what did not so you can make appropriate improvements for the full deployment.

This debrief is done with your Core Team.

7. PROCEED WITH FULL DEPLOYMENT

Critical Tools

Deployment of a New Tool

Planning Checklist

Deploying a tool is indeed a project. Consequently, ensure that you have a deployment plan, a schedule, budget, etc. Working with your Core Team, or with the peers who have agreed to help, you can develop a tactical plan that will ensure a successful deployment.

The list below is intended to help you formulate your own plan. You may need to add your own items to the list. Also note that the list below is in no particular sequence. Finally, the list is not rigid. Use your judgment and do what makes sense within your organization.

Planning Checklist

- ❑ Schedule for the deployment defined, documented and approved.
- ❑ Training materials for employees developed and documented.
- ❑ Training materials produced — hard and soft copies — available for use.
- ❑ Training materials placed in the intranet or web page.
- ❑ Trainers of the new tool identified and informed of their duties.
- ❑ Training of employees scheduled and employees notified.
- ❑ Training of managers and supervisors scheduled.
- ❑ Feedback mechanism in place.
- ❑ Metrics to measure the progress and eventual success of the deployment identified and ready to use.
- ❑ Coaching and encouragement for managers, employees and stakeholders defined and ready — including those who will do the coaching.
- ❑ Support available; for example, facilitators, coaches, tech support, assistants, etc.
- ❑ Obstacle removal in place.
- ❑ Templates, checklists and other tools ready and available.
- ❑ Information on our web page done.
- ❑ Processes that will be impacted by the deployment identified and corrected.
- ❑ Changes to our quality manual completed.
- ❑ Manuals, guides and other reference materials that will be needed by the users ready and available.
- ❑ Executives briefed and ready to fully support the deployment.
- ❑ Human resources (HR) informed.
- ❑ Training department informed.

Product Development

Deployment of a New Tool

Planning Document

This template may help you secure assistance from your executives. It is also a good point to start the process of deployment. It may even become a standard template for deployment of future tools and methodologies. Modify this template to meet your specific objectives within your organization.

1. Tool or methodology to be deployed.

2. Objective of the tool deployment.
   - What will be the outcome once the tool has been successfully deployed?
   - Why are you deploying this tool?

3. Who is the expert on this tool?
   - Who is the internal person who knows the tool and is willing to provide guidance to others?

4. What documentation is available?
   - Models.
   - Software.
   - Books.

5. Are there any external resources available?
   - Consultants.
   - Instructors.
   - Facilitators.

6. What are the risks involved in the deployment?

7. How will we pilot the tool?
   - Which team, organization or group will be the first to apply the tool to actual work.

8. What will be our communication plan?
   - With executives of the organization.
   - With employees.
   - With the pilot team.
   - With other stakeholders.

9. What will we measure to assure us that the deployment has been successful?

Deployment of a New Tool

Tips for Successful Deployment

There are a myriad of things you must consider before you deploy a new tool in your organization.

Brainstorm with your team to find innovative ways to ensure the success. Your corporate culture, industry and reporting structure will help you determine the best way to embed the new tool.

Key Questions

- How can we make the new tool a habit in our organization?
- How do we turn the novelty of the tool into the normal way we do things?

10 Tips for Successful Deployment

1. START SMALL — PICK A SIMPLE PROJECT FIRST
The intent at this stage is to ensure that your team knows how to use the new tool.

2. CONSIDER THE CULTURE OF THE ORGANIZATION
If your organization has not used this tool before, you need to lead your team to understand and embrace it. Some will push back, and you need a plan to deal with some skepticism and denial.

3. PROVIDE THE RIGHT TRAINING TO THE RIGHT PEOPLE AT THE RIGHT TIME
Your team, and even some managers, may need training. You must provide the skills necessary to succeed.

4. ENSURE CROSS-FUNCTIONAL PARTICIPATION
Most tools provide maximum benefit when people from several functions participate. Moreover, most projects involve people from several areas of the organization, and they must be included.

5. PUT IN PLACE A PROACTIVE FEEDBACK MECHANISM
Enable your team to provide you with input on the new tool.

6. BE AVAILABLE TO REMOVE OBSTACLES
Inevitably, your team will run into obstacles when implementing a new tool for the first time. Put in place a mechanism that will enable you to remove these obstacles; otherwise, the team may reject the tool altogether.

7. DEVELOP AN ESCALATION STRATEGY
Some of the issues and challenges that will emerge in deploying your new tool will need the attention of your managers and other executives. Plan an escalation process before things come to a boil.

8. MAKE INFORMATION ABUNDANTLY AVAILABLE
When first deploying a tool, your team, managers and executives will need information. Make certain that information is abundant and conveniently available to all.

9. MAKE COACHING AVAILABLE
Your team members will need one-on-one assistance. Even if you provide training, individuals will have questions and other barriers that you must help remove.

10. CELEBRATE
Recognize progress — celebrate small and large victories. Everyone needs reassurance that progress is being made.

Product Development

Yellow Sticky Protocol

Effective Group Participation

Benefits

The ability to ensure that everyone in a group participates is greatly enhanced through the use of Post-It Notes® (trademark of 3M), known in the vernacular as "yellow stickies." Another benefit of yellow stickies is that they provide a self-documenting process; rather than writing things on a flip chart, which is time-consuming, yellow stickies significantly accelerate the documentation. Finally, yellow stickies are portable; that is, one can move them, hand them over, rewrite them, etc.

Some individuals, particularly senior executives, are not fond of yellow stickies. This is due to a misunderstanding as to who uses them. But this is a serious tool that will produce better results, and more quickly. These people need to be reassured that, barring a technological breakthrough, yellow stickies are still a fine tool that everyone, including executives, can use.

The process favored in this guidebook relies heavily on the proper use of yellow stickies.

Establish the Protocol

To derive the benefits of using yellow stickies, the facilitator must establish a protocol — an articulated set of guidelines as to the way the information will be presented in each yellow sticky.

"The process favored in this guidebook relies heavily on the proper use of yellow stickies."

The protocol is a way to create a safe environment. For example, use the protocol to assuage members who may have reservations about their poor spelling, grammar or handwriting.

The protocol is a way to choreograph the process so it yields maximum benefits.

For the sake of productivity, let alone sanity, the protocol must be enforced. All team members must participate and follow the protocol. Every yellow sticky must conform to a standard to be useful.

There is flexibility, so you need not do it exactly as in the corresponding detailed process. The overall goal is to be able to track and process a large amount of information.

Your team must agree on the way the yellow stickies will be structured. Once you have reviewed and highlighted all of your notes, you proceed to transfer the information to the yellow stickies. You need a protocol as to the type of information and its location. This is critical to provide continuity, traceability and consistency

Plan Ahead

Schedule a three to five hour meeting for a team of up to 15 people in a large conference room. Prepare topics and objectives for the meeting as well as the introduction. Gather materials:

- Post-It Notes® (3M brand, model #655 3x5).
- Sharpie® (Sanford brand, fine point, black markers) for all attendees.
- Flip chart paper and easels — two are best.

For comments, questions or additional copies please e-mail the author: debriefs@rapidinnovation.com.

Yellow Sticky Protocol

Critical Tools

Level 3 Detailed Process: Yellow Sticky Protocol

1

PREPARE YOUR MATERIALS

- 3x5, 3M-brand, Post-It Notes® (#655) — One pad per participant. Consider that several of the participants will have a sizeable number of yellow stickies — in some cases as many as several hundred.
- Sanford-brand Sharpie® markers, fine point, black.
- Flip charts or a place to hang the yellow stickies once filled.

 You also will need the participation of everyone who made customer visits.

2

MAKE THE ANNOUNCEMENT IN FRONT OF THE GROUP

Face the audience, make eye contact and say...

"We are going to use yellow stickies in this meeting.

There are three key reasons why this is important:

- It is a self-documenting process, so it saves time and allows for documentation after the meeting.
- It allows everyone to participate.
- We can move the yellow stickies around depending on our needs — no need to rewrite, just move them.

 Caution: Some individuals, particularly senior executives, are not fond of yellow stickies. This is due to a misunderstanding as to who uses them. But this is a serious tool that will produce better results, and more quickly. These people need to be reassured that yellow stickies are a fine tool that everyone can use.

3

EXPLAIN THE PROTOCOL TO THE GROUP

"In order to take full advantage of the tool, we must follow a simple set of rules..."

For example:

1. Don't worry about spelling or grammar; no one will check it. This goes for everyone in this room.
2. Don't write on the sticky side of the note.
3. Write landscape (horizontally) oriented.
4. Use your Sharpie® marker. This will force you to write larger than usual and more legibly, and it will enable others to read notes from a distance.
5. If you make a mistake or change your mind, simply toss it and start a new one.
6. Write only one item per yellow sticky. Only one question, one idea, one quote.

 Encourage everyone to write legibly. Ask them to print. Many of us do not have the best of penmanship. Explain to the participants that they are writing for others to read! It will save time and reduce confusion.

4

START THE PROCESS

 See Also: Affinity Diagrams

48 For comments, questions or additional copies please e-mail the author: debriefs@rapidinnovation.com.

Product Development

Yellow Sticky Protocol

Example: Yellow Sticky Protocol

The topic of conversation from your discussion guide.

This must match the topics contained in the discussion guide that was created for this set of customer interviews.

An agreed-to indication of the importance, relevance or some other priority. In this example, "H" indicates "High" passion or importance of the statement from the source, whereas "M" would indicate "Medium" and "L" would indicate "Low."

Your team can develop similar priorities, depending on the objectives of your visits. Having a priority set will save considerable time during the later stages of this process. Keep it simple; otherwise, your team members will be confused and misinterpret the code.

The Protocol

This area is reserved for the message from the interviewee — the voice of the customer.

It can be a quotation, a fact, a finding, information or any sentence that is relevant. This is extracted from the notes you took during the interview — you must be able to trace it to your notes.

> Topic   H   Customer's title
>
> "Actual quote from the customer, extracted from the interview notes."
>
> Your name/initials   Page, Interview #

The upper right corner is reserved for the title or function of the person who was interviewed.

You need to develop codes to be used here to help you understand the context, particularly if your team interviewed customers from different functional groups.

The name of the customer is not relevant.

Quote marks (" ") are reserved for actual customer quotes! This is a critical rule.

Quotes are always verbatim messages from the customer. Do not use quotes if the customer did not actually say it.

Example

> Ease of use   H   Design engineer
>
> "More than three layers of menus make products difficult to use."
>
> Fred T   Page 1, 2nd interview

A simple code will enable you to trace the information to your notes, and thus to an actual interview.

It is difficult to remember who said what. During the processing and various meetings, you will want to refer to your notes to clarify the context of the statement.

This is simply a tracking mechanism. You should be able to track back to the actual source of the information.

Yellow stickies are landscape (horizontally) oriented)

This example shows a super set of the Yellow Sticky Protocol. You need not use all of the items, just those that will enable you to track and code the critical information. When you have several hundred yellow stickies, you will rely heavily on the protocol to make sense of it all.

For comments, questions or additional copies please e-mail the author: debriefs@rapidinnovation.com.

Yellow Sticky Protocol

Examples: From Requirements to Features and Attributes

Example 1: The Process of Imaging

 The example below is for the purpose of illustration. Please don't focus on the content, but rather the structure and the intent.

Easy to carry around — Businesswoman

The laptop I have is not convenient. I carry it all day from customer to customer in its carry-on bag. But I also have to carry the electronic projector and my business files. When the day is over, my shoulder hurts.

> The three headers (categories) are combined into a single image (message). The image is a customer quote that captures the key message from the customer.

> Headers from the Affinity Diagram capture the key message of the clustered yellow stickies.

Easy on the shoulders

The laptop shall be easy to carry by an average businesswoman

The user shall be able to go places carrying all the required equipment using only one hand

---

Easy on shoulders — Busnwmn. 42y.
Once you add the accessories, the additional battery and other stuffs you get easily to 7 lbs. That's too much to carry all day.

Easy to carry — Insurance agent 52y.
The bag I have is bulky. Why don't they make slim ones? That's easier to carry.

Convenience — Saleswoman 35y.
I do not need one bag for each thing I need to carry. Carrying more than one bag is a pain. Although sometimes I only need the laptop.

> Adhere to your Yellow Sticky Protocol. In this case, the upper right corner captures the source of the quote or information, and everyone in the team must use the same corner for the source.

Easy on shoulders — Saleswoman 35y.
I do not know how much it should weigh, but I know that I can carry a few books without too much pain.

Easy to carry — Businesswoman 42y.
When I travel by air, I have one of these suitcases with wheels. That's the most important woman-friendly innovation of the 20th century. (Laughter)

Convenience — Insurance agent 52y.
I'm always worried if I have all the needed accessories. I stop in so many places that I always run the risk to forget something.

Easy on shoulders — Saleswoman 35y.
These shoulder straps are hurting. You don't see kids carrying their bag this way anymore. If it's not good for kids, why is it for us?

Easy to carry — Saleswoman 35y.
I am not tall (5'). These laptops are made for tall men.

Convenience — Saleswoman 35y.
I do not like arriving and being unable to greet a customer because I have to carry three bags.

Product Development

Yellow Sticky Protocol

Example 2: The Process of Imaging

Ease of Use — Design Engineer
"When I travel for a day or two, I carry my business documents and personal belongings plus my laptop. I wish I had one item that I could carry on board that would save me time and aggravation."
Fred T.

> In this example, the upper left corner captures the topic from the discussion that gave rise to the quote or information.
>
> The upper right corner captures the title of the person who was interviewed.
>
> The lower left corner captures the name of the team member who wrote the yellow sticky.
>
> Every team member must use the same protocol.

The user shall be able to carry business files, laptop and personal belongings in a single carry-on bag.

The bag shall withstand regular airline transportation "mistreatments"

The bag shall allow easy access to the three types of goods that it contains. In particular, the laptop shall be easily pulled out from the luggage compartment.

All in one — Design Engineer
I can only take one carry-on piece in the cabin on many flights.
Fred T

Sturdy — Technician
I'm always anxious to see if my bags were damaged during travel — that's why I don't like to check them in.
MC

Convenience — Design Eng
I have to take the bag out of the luggage compartment to take out my laptop.
Fred T

All in one — Eng. Manager
I hate to wait for my bags when I arrive at my destination. I always wonder if they are going to arrive with me, and sometimes they don't.
Roger F.

Sturdy — Technician
My laptop needs to be protected inside the bag. The last thing I want to see is a damaged laptop when I get to my destination.
MC

Convenience — Technician
I like the wheels and the handle that my carry-on luggage has.
MC

All in one — Technician
I attached the laptop bag to my carry-on case to get past the limit. Sometimes they let me get away with it, but if not, I have to check it in.
Fred T

Sturdy — Mechanical Eng.
The last piece of luggage I had did not even last two years. The fabric was torn in several spots.
Roger F.

Convenience — Manager
I don't want to put everything together in a single bag. I want to have separate compartments to pack various items.
Fred T

All in one — Technician
Registering luggage and waiting for it adds precious time to my trip. I could do without this burden
Fred T

Convenience — Manager
When I go on a business trip, I take my laptop and all my business files with me.
Fred T

Yellow Sticky Protocol

Critical Tools

Example 3: The Process of Imaging

[Pink sticky - top]
Easy to carry around Businesswoman

The laptop I have is not convenient. I carry it all day from customer to customer in its carry-on bag. But I also have to carry the electronic projector and my business files. When the day is over, my shoulder hurts.

[Blue header stickies]
- Easy to Learn
- Easy to Use
- Security

Easy to Learn column (yellow stickies):

Intuitive (easy to learn)
It does not follow the generally accepted practices for designating menus such as Microsoft. I had to relearn a new GUI.

Intuitive (easy to learn)
The Help menu is of little help. It does not allow me to solve a problem — just shows how to use each function.

Intuitive (easy to learn)
I do not want to have to read a manual to start operating new software. I want it up and running in a few minutes. Later, when I get a bit more experience, I might want to look in the manual.

Intuitive (easy to learn)
When I received the package, I looked for a quick-step guide. I could not find one. I had to go through the lengthy explanations in the manual.

Easy to Use column (yellow stickies):

Easy to use
It does things on its own that I don't understand, and don't know how to avoid. For example...

Easy to use
Once I've learned a new product, I do expect to use it without surprise. Your is full of surprises.

Easy to use
Exporting a file to MS Word takes four steps. It is a task I (and I suppose others) do frequently.

Easy to use
I get at least two error messages per hour of operation. And usually, I have no clue what I did wrong.

Security column (yellow stickies):

Security
It does not seem to be compatible with my firewall.

Security
Setting the four Privacy parameters is a chore. You should have a Wizard.

Security
It does not seem to be compatible with my Norton firewall.

Security
It slows down significantly when set at maximum protection. That is just not acceptable, and in addition it is not even mentioned anywhere.

[Callout box]
In this example, two similar yellow stickies appear in one cluster, which is acceptable as long as it makes sense to your team.

For example, use duplicate yellow stickies for emphasis or because, although similar, they convey different messages.

---

52 For comments, questions or additional copies please e-mail the author: debriefs@rapidinnovation.com

Product Development

Affinity Diagrams

Organize Information and Reach Consensus

Introduction

Affinity Diagrams are a very practical, easy to use method that can be applied in many areas once you learn how to facilitate the process — you lead your team through the process, and your team members look to you for instructions and guidance to reach the results.

Affinity Diagrams can be used in many areas of team work:

- Prioritize topics from a list of many
- Collaborate to solve a problem
- Develop a strategy
- Define group options and make choices
- Identify alternative paths or options
- Brainstorm discuss a set of issue

In short, Affinity Diagrams allow you to tap the power of a team by enabling everyone to participate and ensuring that everyone has the opportunity to innovate and create.

Affinity — A natural liking for or inclination toward somebody or something or a feeling of identification with somebody or something.

Affinity Diagrams — A tool to organize information and reach consensus."

"Affinity Diagrams are essential to the process favored by this guidebook."

Historical Background

Affinity Diagrams belong to a set of tools called the KJ method, which is named after Kawakita Jiro, the Japanese anthropologist who first introduced the approach in 1967. The approach has its roots in the values of Zen Buddhism.

Mr. Kawakita developed the method as a means of organizing diverse observations and qualitative information into useful, documented facts. He observed that humans are not linear thinkers; that is, ideas flow in a non-linear fashion. Further, humans tend to jump around when brainstorming. Thus, trying to structure brainstorming sub-optimizes the creative flow.

Why Affinity Diagrams

- They are simple and non-threatening. As the saying goes, "They're not rocket science." This creates an inviting environment for full participation.
- They are inclusive. Properly used, an Affinity Diagram allows everyone on a team to fully participate, regardless of stature or pecking order.
- They are a fine way to ensure consensus in a group, as well as to build the sense of team.
- They are a practical way of organizing seemingly unrelated facts.
- They are very conducive to the creation of clarity and communication within the team.

General Recommendations for Affinity Diagramming

1. Planning your session is essential to success. Please do not treat an Affinity Diagram process as a casual affair. Consult the process map in this document for additional information.
2. Select a room large enough to allow the team to move around. You may have to move furniture to create a clean space to work.

For comments, questions or additional copies please e-mail the author: debriefs@rapidinnovation.com.

Affinity Diagrams

3. Affinity Diagrams work best when done by a group of about 10 people. Larger groups tend to fracture, meaning that some tend to marginalize themselves and not participate. Thus, groups larger than 10 require considerable facilitation and constant focus to ensure productivity.

4. Silence or talking? The general rules indicate the whole process should be done in silence to allow creativity to flow and avoid any one individual influencing the group. We believe that, in certain circumstances, talking is desirable. When you have a small group, say about three to four people, or when everyone is a peer and collaborative by nature, these are examples of when talking might help. Use your judgment and decide based on the dynamics of your group.

5. Use the traditional rules of brainstorming — for example, do not allow criticism of ideas, encourage participation by all team members and generate the largest possible number of ideas, including duplicates and repetition.

6. Remember the follow-up. Many fine sessions end up delivering little or no value due to the lack of action after the session. Ensure before you start that you are committed to following up well after the session is completed. Every session must have an action plan ("road map") to ensure that you produce tangible results.

7. It is best to use flip chart paper. We also recommend other types of large-format paper such as butcher paper, newsprint rolls and _. Do not hesitate to create a large space with paper. Remember that your team will be working in this area — six to nine horizontal linear feet of paper is best.

8. Avoid cryptic descriptions, such as one-word statements like "energy" or "easy." Encourage team members to use the proactive or active tense — for example, "There is not enough energy" or "Make it easy to use." Therefore, statements should have a noun and a verb to avoid ambiguity and enhance clarity.

9. The number of categories or columns should be kept to about five; more than that confuses the teams.

10. Create the feeding frenzy. It is fun and productive to encourage the team to generate as many ideas as possible. Set challenges — for example, ask everyone to generate five ideas. Do many rounds, create contests and give a prize to the person who generates the most ideas.

High-Level Process

1 START
Deliver instructions.

2 BRAINSTORM
Participants create yellow stickies.

3 POST
Place yellow stickies on the wall.

4 ORGANIZE
Participants organize yellow stickies into categories.

5 PRIORITIZE
Participants prioritize categories.

6 DEVELOP ACTION
Assign responsibilities and develop a timeline.

 Important This is a high-level view of the process for an Affinity Diagram. It will help you navigate, without the complexity of all the details; i.e., the Detailed Process Map.

Product Development

Affinity Diagrams

Detailed Process: Affinity Diagrams

INVITE THE PARTICIPANTS

Who should attend?
- All stakeholders in the process or project.
- Customers, if appropriate.
- Try to keep the group to 10 participants or fewer.

PREPARE
- A large conference room
- Post-It Notes® (3M brand, model #655 3x5)
- Sharpie® (Sanford brand, fine point, black markers) for all attendees
- Flip chart paper and easels — two are best

See Also: **Yellow Sticky Protocol**

SET UP THE TOPIC
- State the topic in the form of a question. Ensure that everyone understands it. You may want to discuss it with the group.
- Reassure everyone that there are no wrong answers.
- Reassure everyone that grammar and spelling are not important.
- Explain how to use the yellow stickies — one per subject, use a Sharpie® pen, landscape (horizontally) oriented.
- No criticism.

START THE PROCESS
- Give the question or topic, then ask everyone to write at least three yellow stickies and wait.
- Ask each person to read his yellow stickies as loudly as possible.
- Tell everyone to listen intently as others read their yellow stickies.
- Encourage participants to write additional yellow stickies as others present theirs.
- Collect all yellow stickies and place them on the flip chart or board at random.
- Continue until the group runs out of ideas or repetition sets in.

SCRUBBING THE AFFINITY DIAGRAM

Clean up your Affinity Diagram once all the yellow stickies are written and placed on the board.
- Eliminate yellow stickies that are obviously irrelevant or trivial.
- Eliminate obvious duplications.
- Rewrite yellow stickies that are difficult to read.
- Rewrite yellow stickies that are not clear.

Important: For any and all changes you make when scrubbing the Affinity Diagram, make them with the agreement of the authors.

CONTINUED

55

For comments, questions or additional copies please e-mail the author: debriefs@rapidinnovation.com.

Affinity Diagrams

Critical Tools

CONT'D

6
START THE AFFINITY PROCESS

Rally the group to move up around the flip charts with the yellow stickies.

Ask participants to observe how the various yellow stickies cluster or group into subjects.

Ask the participants to start moving the yellow stickies, using their own criteria to cluster them, regardless of how others feel about it.

Do this in silence! No one can talk during this step!

Stop when there is little or no movement of yellow stickies.

7
PICK 'N CHOOSE

Review each column or cluster that was created in the previous step, with participation by the group.

Ask the group to explain why the cluster or column exists. "Why are these yellow stickies together?"

Move the yellow stickies that do not belong in the column to a different one or a separate place on the flip chart.

Eliminate duplicates by choosing one that represents the topic and placing the others underneath.

For each column, assign a header that captures the topic. Write the header on a sticky notes, preferably a color other than yellow.

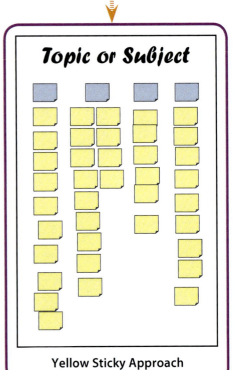

Yellow Sticky Approach

Prioritizing Affinity Diagrams

Karnaugh Maps

8
PRIORITIZE

Once the notes are prioritized, you can begin to assign responsibilities and action items as appropriate.

Also, develop a timeline. This is a critical step!

Save the flip charts for archival purposes or to be transcribed if necessary.

Debrief the group on the effectiveness of the tool and ways to improve the process.

56 For comments, questions or additional copies please e-mail the author: debriefs@rapidinnovation.com.

Product Development

Prioritizing Affinity Diagrams

The Process to Prioritize

Most often, Affinity Diagrams are used to prioritize a set of ideas, choices, things, etc. The criteria for prioritizing are determined by the initial objective that you set for the Affinity Diagram — for example, you might want to do an Affinity Diagram to assign responsibilities or action items. You might also want to pick a few from a selection of many. In short, use the same objective you set for the Affinity Diagram to select the method you will use to prioritize.

Group Consensus

Prioritizing is all about making decisions; keep this in mind as you work with your team. You want to reach consensus. 100 percent agreement is desirable, but not essential. Most of your prioritizing efforts will not have 100 percent consensus. That being said, prioritizing an Affinity Diagram is a team effort.

Select The Criteria

The key to prioritizing is the selection of the criteria used to make the decisions. Take your time, discuss with your team, and above all ensure absolute clarity. Criteria should have two or three elements; in a team environment and without computers it's very difficult to keep track of more than three criteria.

This process should be driven by the strategic importance of the project. For small, simple projects the criteria might be somewhat subjective. If the project is critical, then the criteria must be defensible — objective, verifiable and in full alignment with the strategic goals of your project. This is a qualitative exercise, so avoid criteria that are too complex.

Common Methods

1. ABOVE OR BELOW THE LINE

Draw a horizontal line at the midway, and ask the team to decide for each yellow sticky if it should be above or below the line. This is the quickest approach and one that most people find intuitive. You can prioritize the yellow stickies or the headers depending on your objectives.

2. 1 ROUND OF DOTS

This is a popular method because it's simple and familiar. Give everybody five to 10 self-stick colored dots to place on the yellow stickies based on the predetermined criteria. They can place their dots anywhere, as many dots as they wish per item. When done, simply count the dots to establish priority.

3. 2 ROUNDS OF DOTS

A variation on this theme is to do two rounds of dot placing with five dots for both rounds:

> Round 1: Place five dots anywhere, with no guidelines.
>
> Round 2: Place five dots, but only on two yellow stickies.

4. 3 ROUNDS OF DOTS

A more sophisticated method involves giving each team member 30 dots to be placed during three rounds:

> Round 1: Place 10 dots anywhere, with no guidelines.
>
> Round 2: Place 10 dots, but no more than two dots on any one yellow sticky.
>
> Round 3: Place the remaining 10 dots, but only on two yellow stickies (five and five).

5. YOUR INVENTION

Invent your own system for prioritizing. As long as it's rational and leads to clear decisions, you'll be fine.

For comments, questions or additional copies please e-mail the author: debriefs@rapidinnovation.com.

Critical Tools

Prioritizing Affinity Diagrams

Level 3 Detailed Process: Prioritizing Affinity Diagrams

① SELECT THE CRITERIA

The set of criteria is generally two to three elements that will determine priority. The criteria will drive the decision process.

Take your time, discuss with your team, and above all ensure absolute clarity.

The simplicity or complexity of your criteria depends on the strategic importance of the project. As a rule, keep it simple.

Once agreement has been reached, write the criteria on a flip chart to help the team stay consistent.

 This process should be driven by the strategic importance of the project. For small, simple projects the criteria might be somewhat subjective.

If the project is critical, then the criteria must be defensible — objective, verifiable and in full alignment with the strategic goals of your project.

② START THE PROCESS TO PRIORITIZE

Q. Do we prioritize the headers, or each yellow sticky?

A. It depends on your objectives:

The preferred method is to prioritize the individual yellow stickies. Based on which categories the prioritized yellow stickies are in, you can infer the priority for the categories.

For example, if you use one of the Dots methods, count the number of dots in each category, and write it on the header. This will create an overall priority ranking.

Another approach is to ask the participants to place the dots on the actual headers — not the yellow stickies. This is faster, but it may miss an important message: Which yellow sticky is most important?

Use your judgment as to which voting method will serve you best.

 Prioritizing is a team effort. 100 percent agreement is desirable, but neither essential nor expected. The goal is consensus.

③ PRIORITIZE

Choose from one of these common methods or invent your own:

- **ABOVE OR BELOW THE LINE:** Draw a horizontal line, and ask the team to decide for each yellow sticky if it should be above or below the line.
- **1 ROUND OF DOTS:** Give everybody five to 10 self-stick colored dots to place on the yellow stickies based on the criteria. They can place their dots anywhere, as many dots as they wish per item. When done, simply count the dots to establish priority.
- **2 ROUNDS OF DOTS:** A variation on this theme is to do two rounds of dot placing with five dots for both rounds:
  - Round 1: Place five dots anywhere, with no guidelines.
  - Round 2: Place five dots, but only on two yellow stickies.
- **3 ROUNDS OF DOTS:** A more sophisticated method involves giving each team member 30 dots to be placed during three rounds:
  - Round 1: Place 10 dots anywhere, with no guidelines.
  - Round 2: Place 10 dots, but no more than two dots on any one yellow sticky.
  - Round 3: Place the remaining 10 dots, but only on two yellow stickies (five and five)

Product Development

Prioritizing Affinity Diagrams

Example: Affinity Diagram

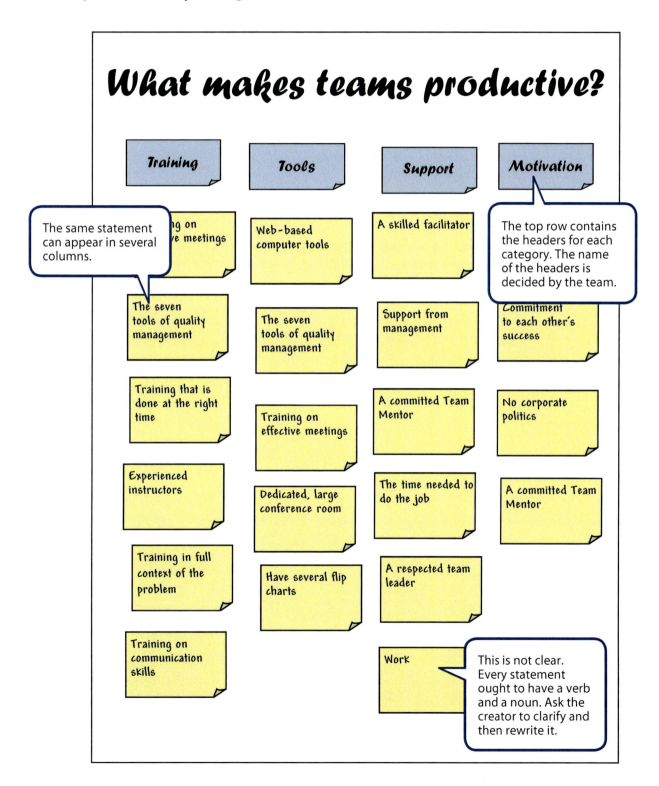

Critical Tools

Problem-Solving Process

Problem Solving Process

Introduction

Program managers spend considerable time solving a stream of problems due to the fluid nature of technology. Many of these challenges are straightforward and can be readily dispatched.

However, sometimes tough problems are under-managed. Organizations wind up drifting, unsure of the reasons for change, the severity of the problem or the path forward.

A disciplined problem solving approach should be applied for the most difficult crises that will shape the direction of the business.

General Rules for Solving Problems

- Do not underestimate the need to apply a formal process to solving a problem. Some in the team might dismiss it as trivial or something that can easily be resolved.

- Beware of the easy solutions. Many times the problem is a symptom of a much larger issue. There is always the desire to jump to the solution without the process to diagnose the real root of the problem. Here is where a formal process of solving problems plays a very important role.

- Widely communicate the solution. Frequently a problem has been solved, but the majority of the team does not know it. The process for communicating the disposition of a problem should be as rigorous as the process to solve it.

Overview of the Problem Solving Process

1. Identify the problem. Come to a common agreement on the nature and extent of the issue.

2. Generate criteria for a solution. Select which business measures will guide the decision process.

3. Generate alternative solutions and test them against the business criteria you have selected.

4. Conduct diligent reviews on key options. Prune the options to the few that merit more detailed analysis against the criteria.

5. Decide! Make a clear decision and then back that decision with a detailed implementation plan.

6. Communicate the solution to all stakeholders.

7. Learn from the process

TIPS is the acronym for Theory of Inventive Problem Solving, and TRIZ is the acronym for the same phrase in Russian.

TRIZ was first developed in 1946 by Genrich Altshuller and his colleagues in the former USSR and is now being developed and practiced throughout the world.

TRIZ research began with the hypothesis that there are universal principles of invention that are the basis for creative innovations that advance technology. Furthermore, if these principles could be identified and codified, they could be taught to people to make the process of invention more predictable. The research has proceeded in several stages over the last 50 years. More than two million patents have been examined, classified by their level of inventiveness and analyzed to look for principles of innovation. The three primary findings of this research are as follows:

Problems and solutions were repeated across industries and sciences.

Patterns of technical evolution were repeated across industries and sciences.

Innovations used scientific effects outside the field where they were developed.

For additional details visit: http://www.triz-journal.com.

Product Development

Problem-Solving Process

Detailed Process: Problem Solving

PREPARE FOR YOUR PROBLEM SOLVING MEETING

Being well-prepared will make your meeting more successful and quicker:

- Gather all materials that will help you add clarity to the process; for example, technical documentation, reports and drawings.
- Select the attendees based on their ability to help solve the problem. Do not hesitate to invite people who are not directly involved with the problem but can add value to it. For example, experts on the topic and outside consultants.
- Ensure diversity of participants; the more points of view the better. For example, combine people who are involved in the problem with those who are not. People from other functions may also be helpful, as will some of your suppliers and vendors.
- Take care of all logistics; for example, reserve a venue that is large enough for your team to work.

START YOUR PROBLEM SOLVING MEETING

Share the perceptions of the problem: How does each team member see the problem? This is a critical step, as the term "problem" may mean different things to different people. Spend enough time in this definition phase — don't shortchange it.

Write the various perceptions on the flip chart. You can also use yellow stickies.

Clarify the definition of the problem by agreeing on a single wording of the problem.

Identify and document on a flip chart the expectations for a solution from your team members. As with the definition of the problem, the perception of a solution will vary. It is important to reconcile; i.e., find a common purpose.

> **Important!** We have various ways to identify something that is not desirable; for example, we might call it a flaw, failure or fiasco.
>
> Some of these terms carry a certain stigma or prejudice. Some cases imply punishment.
>
> It is important that your team agrees on a term that is not politically loaded, but rather brings objectivity to your dialog. You cannot afford distractions.

CONFIRM PROBLEM STATEMENT

At this point you will have spent a considerable amount of time writing and re-writing the problem statement.

Write the problem statement one last time, as clear and readable as possible.

Validate with your team by asking everyone to agree or disagree with the statement. Does it truly represent the problem to be solved?

This is the most critical step of this process, the clear articulation of the problem, plus the support of the entire team.

You may need to spend additional time refining the problems statement, which would be time well-spent!

Do not proceed until there is clear alignment.

> **Important!** The problem statement — the one you and your team have crafted — will most likely be the hardest step in the entire process.
>
> Don't hesitate to rewrite the problem statement several times; it's worth the effort.
>
> Spend all the time you need to ensure clarity and consensus among team members.
>
> Do not proceed until everyone understands the statement of the problem.

CONTINUED

For comments, questions or additional copies please e-mail the author: debriefs@rapidinnovation.com.

Problem-Solving Process

Critical Tools

CONT'D

4 PROFILE THE PLAYING FIELD

Analyze the problem based on the:
- **GIVENS:** Things that cannot be changed as part of the solution to the problem such as raising prices far in excess of what the customers are willing to pay.
- **FORCES AT WORK:** Those that are acting for and against the solution. Also known as Force Field Analysis.
- **MUST-HAVES:** Things that are non-negotiable or cannot be changed. Very similar to the givens, but with a focus on unacceptable trade-offs; for example, a reduction in product performance.
- **DESIRABLES:** Things that would be nice to have as part of your solution.

If appropriate, review the initial objectives of the project.

The outcome of this section is a single definition of the problem — the problem statement — and a set of criteria that bind the problem (a combination of steps 1 and 2).

5 DEFINE AND DOCUMENT SOLUTION CRITERIA

- Describe an effective solution (use the Yellow Sticky Protocol).
- Identify metrics to measure success of solution.
- Identify and list costs, customer needs and resources.
- Define the must-have criteria.

6 GENERATE ALTERNATIVE SOLUTIONS

- These are possible solutions, not necessarily real ones. This is a critical step in order to tap the creativity of your team.
- The rules of brainstorming apply, such as don't criticism and keep an open mind. This is not the time to eliminate or stifle ideas.
- Use a brainstorming tool like an Affinity Diagram.

Important We recommend using yellow stickies for the step of generating alternative solutions.

Ask your team to write down every possible solutions to the problem. Then ask your team to read each suggestion before posting every yellow sticky on the wall.

You can use the Affinity Process to organize and prioritize the yellow stickies.

7 EVALUATE ALTERNATIVES AGAINST THE CRITERIA

- Evaluate the proposed solutions (step 3) against the developed criteria (step 2).
- Retest if you have other alternatives to consider.
- Don't underestimate the impact of varying skill levels and cultural gaps.
- Develop a matrix and consider combining alternatives.

Note: You may need to do research in order to validate that the alternatives are viable and achievable. This may take time, even require another meeting. Some teams opt to simply select the alternative, assign responsibility and proceed without validation.

See Also **Critical Tools**

CONTINUED

Product Development

Problem-Solving Process

Cont'd

⑤ Confirm Your Selection of Solutions
- Ensure clear understanding and alignment behind the preferred alternatives and the criteria.
- Conduct due diligence on the top options.
- Complete the fact-finding.
- Document the key assumptions.
- Document risks.
- Check for unintended impacts.

⑥ Develop Your Action Plan
- Develop a detailed plan to verify or validate the assumptions.
- Develop a plan to address the identified risks and uncertainty.
- Establish ground rules for the implementation of the solution. For example:
    - How are you going to deal with changes?
    - How will you monitor progress? Will you track the metrics, milestones or compliance?
    - When and how will you decide that the problem has been solved?
    - Who will be responsible for implementation?
    - Who will be responsible for closure?
    - Who, when and how will you declare victory?
- Close any gaps concerning resources, skills and culture (a gap is a misalignment between "what it is" and "what it needs to be"; for example, you need more resources that you currently have).
- Communicate the decision to all stakeholders.
- This step is all about ensuring execution; that is, the implementation of a road map that will guarantee the solution of the problem. This is a high-risk area, as many solutions fail to be implemented due to the lack of a plan of action.

Learn and Improve (Kaizen)
Your risk management process will improve significantly if you constantly add the lessons you've learned.
- Debrief the team on the process used to solve the problem.
- Identify areas of improvement.

For comments, questions or additional copies please e-mail the author: debriefs@rapidinnovation.com.

Problem-Solving Process

Problem Solving

Flexibility Matrix for the Problem Solving Method

Strengths	Weaknesses
A traceable path of problem to solution.	Need to watch for hidden boundary conditions.
Ensures a lot of input from everyone.	A solution will be found, but it may be politically incorrect.
Participative; creates ownership and encourages diverse solutions.	It involves hours of work, and it may be tedious.
Appropriate for difficult problems.	
A way to dispose of a problem once and for all.	
A disciplined way to identify the real causes of a problem, not just opinions and anecdotal information.	

Flexibility

- You can vary the size of the team depending on the complexity of the problem.
- You can create additional steps to suit the culture of your organization.
- You can do the process in two or three sessions to avoid fatigue.
- Include the documentation in the archive of your project in case the problem shows up again.

The Problems that Refuse to Die

With some regularity, there are problems and issues that have been solved, yet they persist and continue to distract the organization. This may be due to a lack of proper communication; specifically, the problem was solved but only a few in the organization know it. This can be addressed by a proactive, high-energy communication effort. For example, clearly announce the solution during review meetings and send notices out specifically on the conclusion of the issue. It helps to be emphatic.

The other reason for this type of problem is that some in the organization do not wish it to go away. Keeping the problem alive is one way to fight the chosen solution. One way to deal with this difficulty is to use the formal process outlined in this document, which allows you to cover all the possibilities and overtly deal with and dispose of the problem — this is called an intervention. Simply put, deal with the issue directly.

Other Tools for Solving Problems

There are several tools that provide a fine framework for solving a problem; for example, the Affinity Diagram, the Fishbone Diagram (Ishikawa Diagram), Karnaugh Maps (2X2 matrixes) and various cause-and-effect diagrams. Some of these are covered in this document, and the rest are readily available on the Web and in other handbooks and guides.

More advanced methodologies are also available to tackle very complex or esoteric problems; for example, FMEA and TRIZ. These methodologies are also documented on the Web.

Avail yourself of these tools. We suggest that you become proficient in at least one or two.

Product Development

Karnaugh Maps: 2x2 Matrixes

Prioritize Items from a Selection of Many

Introduction

Karnaugh Maps, also known as Karnaugh Diagrams, have their origins in mathematics and are related to Venn Diagrams. They are especially useful for generating the most optimal Boolean expression. By varying the criteria and the values, you can apply this tool to select or prioritize many choices.

A Karnaugh Map is a 2x2 matrix. But it is also a participative tool that enables a group to prioritize or select items from a group of many, using a set of criteria. It is a simplification of a FMEA. In short, it enables you to make better and faster decisions.

Applications

Facilitators are often faced with helping a team make a decision or a selection to set priorities from an array of many options. For example, they must determine the risk factors of a new deployment, select key features on a new product, determine the importance of tasks or projects and much more.

The scientific community has been successfully dealing with this issue for some time. Over the years, engineers and scientists have developed tools and methods to help them cope with unanticipated failures. They have learned successful ways to anticipate what may go wrong in order to prevent it. From risk analyses to FMEA (Failure Model and Effect Analysis) to design of experiments, these methods, when well-understood and carefully applied, have proved to work.

Generally, when choosing we have two criteria to use — for example, "urgent" and "important." While some things can be very important, they are not necessarily urgent. Thus, the ability of a team to separate all tasks based on the set of criteria will result in a better deployment of limited resources.

Another example is risk selection. Take any project and you can ask your team to identify those things that are likely to go wrong, and the negative impact on the project if they do. These two criteria will help the team focus on those things that are likely to happen, and will cause serious damage to their project. Thus, rather than focusing on all the things that can go wrong, they can focus on the vital few.

Why 2x2 Matrixes

- The 2x2 matrix is attractive because it is graphical and simple and enables group participation.
- It engages everyone on the team, and allows for innovation and breakthrough.
- It can be as rigorous as you need it to be — i.e., the results can be made statistically defensible.
- It is very flexible, from prioritizing to selecting, choosing and more.

General Rules for Karnaugh Maps

- You should first be familiar with the concepts of Affinity Diagrams, or KJ (Kawakita Jiro) method.
- A session to develop a successful 2x2 map will take three to four hours.
- The challenge is to select the right criteria for each of the axes. The Y axis can represent one criterion such as urgency. You can assign values; i.e., a scale, such as 0 to 10.

Facilitator
Anyone who is leading a team to success; this includes project managers, team leaders, program managers, supervisors, leads, etc.

Pair-Wise
A process of comparing entities in pairs to determine preference.

The method is used in statistical analysis, specifically probability.

L.L. Thurstone first introduced this approach in 1927, using pair-wise comparisons for measurement.

For our purposes, a group or team must express a preference such as priority between two mutually distinct alternatives in order to meet an objective.

Karnaugh Maps: 2x2 Matrixes

- The X axis (for example, important) is your second criteria. Likewise, you can assign to it a value.
- Once you have a way to plot items based on the criteria (e.g., importance and urgency) place yellow stickies in the appropriate quadrant. For example, the items in the upper-right corner are more important and more urgent than the others.
- The color of the yellow stickies and their size can also represent additional criteria such as cost and difficulty of implementation. We recommend that you start with just two variables.
- We suggest that you limit your Karnaugh Maps to no more than three variables. Because Karnaugh Maps are a visual tool, it is best to limit the number of variables so as not to confuse the team.

Multiple Variables

Can you use more than three variables? Yes, you can. You can use a weighted-average method through the use of an Excel spreadsheet. You can have as many variables as you wish. But, it is important to show the results in only three variables. There is the tendency to complicate matters beyond what you need. Be careful! Let your objectives determine the number of variables. In our experience, most instances only require two or, at most, three.

The Scales

Generally, a scale of 1 to 10 on both axes is sufficient. It is important that the team agrees with the definition of the scale; i.e., decide what each incremental level means. Some teams prefer to use a 1 to 5 scale, which is also acceptable.

Remember, this approach is **QUALITATIVE**. A **QUANTITATIVE** scale is also acceptable, but we highly recommend that you choose these sparingly. The additional accuracy provided by a quantitative scale, in most cases, is overkill. Use quantitative scales when the topic at hand is critical and of extreme importance to your team or company.

A Better Way to Prioritize Affinity Diagrams

You have an alternative to help you prioritize your Affinity Diagram: You can choose the method outlined in the Affinity section (Prioritizing an Affinity Diagram) or you can choose the Karnaugh Map (KM) approach. Both are valuable and useful, although Karnaugh Maps provide more focus and perhaps a bit more accuracy. Karnaugh Maps also provide a more rational and organized approach than Affinity Diagrams. With Affinity Diagrams, you prioritize using a single variable, but Karnaugh Maps allow for two or more variables. For example, you could set the x and y axes to indicate "cost to develop" and "importance to the customer," and these two variables, called pair-wise, combine more than one criterion. You can even add a third variable; for example, the color of the stickies denotes "difficulty to develop." We suggest that you limit your Karnaugh Maps to no more than three variables. For example, the x axis is "cost to develop," the y axis is "importance to the customer" and the color of the Stickies is "difficulty to develop." In short, because Karnaugh Maps are a visual tool, it is best to limit the number of variable so as not to confuse the team. For examples of other combinations of variables, see examples of pair-wise criteria and their application later in this section.

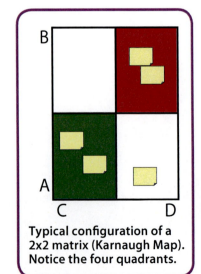

Typical configuration of a 2x2 matrix (Karnaugh Map). Notice the four quadrants.

Breadth of Uses for Karnaugh Maps

Once you learn to use them, you will find that they can be used in areas well beyond customer contacts. The famous 2x2 matrixes have been around for many years. They can be used to prioritize or make decisions. In short, any time a decision needs to be made, you can use Karnaugh Maps.

Product Development

Karnaugh Maps: 2x2 Matrixes

Example: Karnaugh Map (2x2 Matrix)

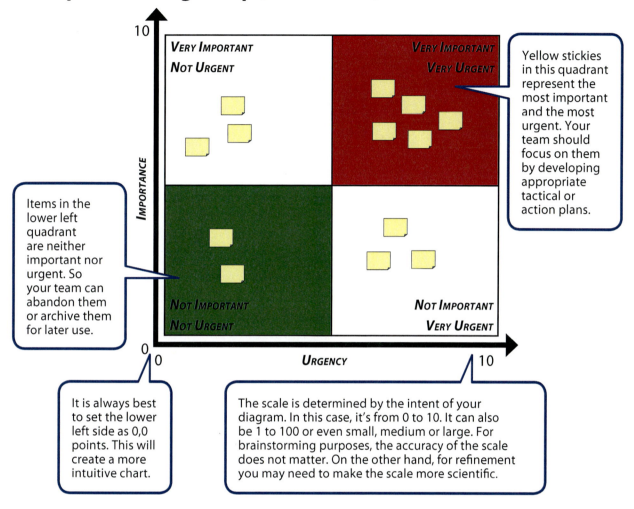

Karnaugh Maps: 2x2 Matrixes

Example 2: Prioritizing Customer Requirements

Deliver the Value to the Customer Quickly

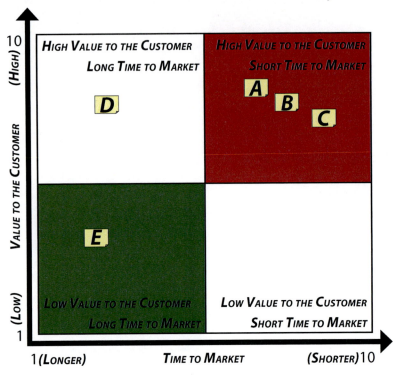

Here's an example of prioritizing based on pair-wise criteria: "Value to the Customer" and "Time to Market."

Items "A", "B" and" C" represent desirable candidates for high priority. Not only would these items deliver the highest value to the customer, but also they take the least amount of time to develop. In contrast, item "E" is not desirable at all.

Deliver the Value to the Customer in a New Way

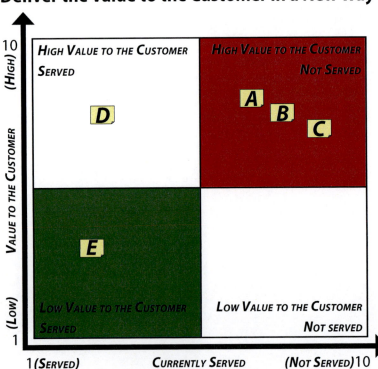

Here's an example of prioritizing based on pair-wise criteria: "Value to the Customer" and "Currently Served."

The variable "Currently Served" indicates how well the customer requirement is being addresses either by your product or by the products from your competitors.

Items "A", "B" and" C" represent desirable candidates for high priority. Not only would these items deliver the highest value to the customer, but also they are requirements that are not currently addressed. Again, item "E" is not desirable at all.

Product Development

Karnaugh Maps: 2x2 Matrixes

Pair-Wise Comparison

Pair-wise comparison generally refers to any process of comparing entities in pairs to judge which of each pair is preferred or has a greater amount of some quantitative property. Pair-wise comparisons are relational — unlike Cartesian X and Y coordinates, where one is the function of the other. The method of pair-wise comparison is used in the scientific study of preferences, attitudes, voting systems, social choice, public choice and multi-agent AI systems. In statistical analysis, it's used to determine probability.

Prominent psychometrician L. L. Thurstone first introduced a scientific approach to using pair-wise comparisons for measurement in 1927, which he referred to as the Law of Comparative Judgment. Thurstone linked this approach to psychophysical theory developed by Ernst Heinrich Weber and Gustav Fechner. In psychology literature, it is often referred to as Paired Comparison.

Application to Product Development

In product development, your team can use pair-wise comparison to express a priority between two mutually distinct alternatives in order to meet an objective.

Many pair-wise criteria are possible. We offer a short list of examples to help you and your team develop criteria that fit your needs exactly.

1. **CHARACTERIZING RISK:** The criteria of "Likelihood of Happening (Probability)" and "Severity if it Does Happen (Impact)" — often expressed as a P-I Factor — maps risks that are most likely to occur and will have the most severity if they do. You can use the results to manage risk.

2. **SELECTION OF OPTIONS:** For example, when selecting projects or courses, use "Attractiveness" and "Cost" to determine best value.

3. **MAKING CHOICES:** When making trade-offs, try "Ease of Use" and "Cost of Implementation." "Achievability" and "Benefits" are simple criteria you could develop to have real meaning — for example "Achievable within the Fiscal Year" and "Financially Beneficial."

4. **DEFINING PRIORITY:** Use "Urgent" and "Important" to prioritize the allocation of resources.

5. **DOING A GAP ANALYSIS:** Use "Current Level versus Desired Level" and "Difficulty in Achieving."

6. **DEFINING COMPLEXITY:** Use "Complexity" and "Difficulty in Achieving."

7. **DOING RESOURCE ALLOCATION:** Compare "People" or "Resources" on the Y-axis with "Priority of Action" on the X-axis. The priority can be defined in strategic terms or some type of urgency scale.

8. **DEFINING THE COMPETITION:** Map the need to defend a segment, abandon it or create a new segment. One example: "Level of Competition" and "Urgency to Defend." Another example: "Level of Competition" and "Critical Market Segment."

9. **ENSURING FIT:** Use "Fit with Business Objectives" and "Cost of Implementation."

Critical Tools

Karnaugh Maps: 2x2 Matrixes

Detailed Process: Prioritizing a Karnaugh Map

SET OBJECTIVES FOR YOUR KARNAUGH MAP

For example:
- Determine risks and rewards.
- Prioritize from a choice of many.
- Select options based on certain criteria.
- Choose a course of action, given certain criteria.

 It's a good idea to prepare a blank 2x2 matrix in advance. Tape two or more sheets of flip chart paper together and use markers to draw the matrix. A white board is an alternative, but upon completion you need to carefully gather the yellow stickies.

 After a brainstorming session, you may end up with as many as 300 yellow stickies. Scrubbing is the step to eliminate duplicates, redundancies and any other non-value-added yellow stickies.

Your goal is to end up with approximately 30 or fewer yellow stickies for the last step. On the other hand, if you end up with many more, and you feel they are important, then proceed.

PROCESS THE INFORMATION

Once all the yellow stickies have been collected, start scrubbing the information to ensure that every yellow sticky communicates a clear, distinct idea.

This step can take 30 to 45 minutes.

If you have too many yellow stickies, then you need to affinitize (group) them into main categories.

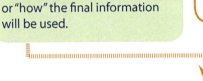 The selection of the two criteria for the X and Y axes is critical. To help determine the criteria, spend sufficient time with the team discussing "why" or "how" the final information will be used.

DEFINE THE CRITERIA AND SCALE

Select the pair-wise criteria for the X and Y axes. For example:
- Important vs. Urgent
- Expense vs. Ease of Implementation
- Risk vs. Rewards
- Important to Customers vs. Expense to Develop

Set your scale. For example:
- Small, medium and high
- 1 to 10 or 1 to 100
- A, B and C

 You can define a third criterion via the color or size of the stickies. Use this option sparingly. Furthermore, when first deploying the Karnaugh Map tool, we recommend you begin with just two criteria.

CONTINUED

Product Development

Karnaugh Maps: 2x2 Matrixes

See Also: Pair-Wise Criteria

CONT'D

4. Plot the Information on the Matrix

Using your criteria, place each yellow sticky inside the 2x2 matrix. Do this in collaboration with your team.

You can use several methods of voting to assign value to each of the yellow stickies — for example, count hands, use dots, assign numerical values to later average. The more critical the topic, the more formal the vote should be.

Don't dispose of the plot, as you may need it for future reference.

5. Reach Your Conclusions

Placing all the yellow stickies does not complete the process!

The value of your 2x2 matrix is in the decisions that you make, the action-items that are generated and the follow through to act on these directives — simply put, the process is not over until something happens.

6. Develop Your Action Plan

Remember, it's not an action plan unless it clearly identifies "do what," "by whom" and "by when."

Document your action plan.

Follow up until all actions have been completed.

For comments, questions or additional copies please e-mail the author: debriefs@rapidinnovation.com.

The 5 Whys

The 5 Whys

Help Customers Reflect Their True Needs

This is a simple methodology to obtain valuable information from customers. As the name implies, in the literal sense, it means asking the customers "why" five times in order to get to the root message — i.e., the most fundamental message concerning the needs of the customer. This methodology originated in Japan and is typically associated with the process of identifying why failures occur. It is part of Root Cause Analysis.

In the context of customer interviews, the 5 Whys serves the purpose of getting past the stock answer. It allows you to peel away the layers of information until you get the valuable information underlying the needs of the customer. The 5 Whys is an excellent tool to get your customers to tell you their unarticulated or unstated needs. The process of asking why causes the customers to reflect on their actual needs.

General Rules for the 5 Whys

1. You can use as many whys as needed. While the name of the methodology is "5 Whys," it simply means that you can continue to ask why until you get useful information. In some cases, three whys will suffice, yet in other cases as many as seven will be needed. Use your judgment, and stop asking when you feel you have obtained the root message or the valuable information you are seeking.

2. Avoid the toddler's curse, when a 3-year-old child drives you crazy asking "why." Instead, let the customer talk, then ask why. You can also use other ways to ask — for example: "Tell me more" and "Why is that?"

3. Help the customer by restating her previous answers or by repeating the questions. For example: "You said that this product is difficult to use, and that this is because the menu system is not intuitive; now, can you describe why the menu system is not intuitive?"

4. The intent of the 5 Whys is to get to a level where you get actionable input and not generalities. Actionable input is customer information that you can actually use in developing requirements and features.

5. The 5 Whys is not intended to be used as an interrogation tool. That is, one does not actually ask "why" five times. We encourage you to use your own style to apply the concept.

6. Your role is to enable your customer to talk and express his "root message."

The 5 Whys

Examples: The 5 Whys

Example 1: The Main Pump Failed

Q: Why did the main pump fail?
A: The coupling from the main intake to the pump broke and created a shutdown of the line.

Q: Why did the coupling from the main intake fail?
A: The "O" ring failed under pressure.

Q: Why did the "O" ring fail under pressure?
A: The night temperature fell to -20 degrees and the "O" ring was not specified for low temperatures.

Q: Why was the "O" ring not specified for low temperatures?
A: No one expected these pumps to be sold in Alaska.

ROOT CAUSE: Insufficient environmental specs to enable safe operation of the pump under arctic conditions.
CORRECTIVE ACTION: Review all environmental specs and redesign for arctic conditions.

Example 2: The Machine Operator Received an Electric Shock when Using the Pick and Place Machine

Q: Why did the machine operator received an electric shock when using the Pick and Place machine?
A: The operator touched the high voltage line.

Q: Why did the operator touch the high voltage line resulting in shock?
A: The safety cover to the high voltage lines was open and the operator reached inside.

Q: Why did the operator reached inside the cover?
A: The operator thought that there were adjustment controls inside the cover.

Q: Why did the operator think that there were adjustments inside the box?
A: The box had no label and was opened.

ROOT CAUSE: Lack of proper safety marking and also a safety box that has no lock nor "Poka Yoke."
CORRECTIVE ACTION: Place warning labels prominently on and about the safety box. Add a lock that only the maintenance technicians can use. Add an automatic disconnect when the box is opened.

The following examples illustrate the concept of the 5 Whys. In both cases there are only three follow-up questions. In some cases, it could take seven or more follow-up questions. The 5 Whys concept can be used in many areas of inquiry.

Critical Tools

The 5 Whys

Example 3-A: Interviewing Customers

Q — **INTERVIEWER:** How do you feel about this software product?
CUSTOMER: It is very difficult to use.

1st Why — **INTERVIEWER:** Tell me more.
CUSTOMER: The menu system is not intuitive.

2nd Why — **INTERVIEWER:** Tell me why it's not intuitive.
CUSTOMER: I get lost all the time.

3rd Why — **INTERVIEWER:** Why do you think you get lost using the menu?
CUSTOMER: Well, there are too many layers and no way to get back.

A — **ROOT MESSAGE:** The menu system has too many layers and no way to get back, which makes customers feel that the product is too difficult to use.

Note the root message — i.e., actionable input from the customer.

Example 3-B: Interviewing Customers

Q — **INTERVIEWER:** Why did the fire start?
CUSTOMER: The problem was a short in the main transformer.

1st Why — **INTERVIEWER:** Why did the transformer short?
CUSTOMER: Water got inside the mechanical package of the transformer.

2nd Why — **INTERVIEWER:** Why did water get inside?
CUSTOMER: The gaskets froze, became brittle and cracks developed, allowing water to seep.

3rd Why — **INTERVIEWER:** Why did the gaskets crack?
CUSTOMER: They are not specified for cold weather.

A — **ROOT MESSAGE:** The gaskets specified in the design document did not meet the environmental conditions where the transformer was placed.

The following examples illustrate the concept of the 5 Whys applied to the important task of interviewing customers

Product Development

The Fishbone Diagram

The Fishbone Diagram

Diagramming Cause and Effect

Generally, when something goes wrong — for example, a machine stops working — we tend to jump to conclusions based on our experiences and immediately infer the reason for the failure. Worse yet, we go all the way to a solution. There is a risk in this behavior. What if the reason for the breakdown is not what we thought? What if our solution is the wrong one?

The Fishbone Diagram, also known as the Ishikawa Diagram, is a fine tool for providing the discipline to go from "symptom" to actual "root cause" of the problem, regardless of the nature of the problem.

Teams are frequently confronted with solving problems; i.e., addressing a snag or an obstacle. Furthermore, the definition of the problem tends to be, in reality, a symptom of something else. For example, "the press does not work" or "the car does not start."

It is very tempting to try to solve the symptom of the problem without trying to analyze the real cause. In the example above, "the press does not work" may lead you to try several fixes, only to find out that the problem persists. Instead, taking time to characterize and prioritize the causes of the problem leads the team to find a permanent solution. This is where the Fishbone Diagram comes into play.

The cause-and-effect diagram is used to discover all the possible causes of a problem; those things that cause the effect; i.e., the symptoms. Further, it helps to prioritize the causes and drive a methodical identification of the true causes of a problem, thus leading to a permanent solution.

Why Cause and Effect

- You can't afford the trial and error to fix problems in your organization.
- It is a well-developed, mature process. It has been shown to be reliable and very practical.
- It taps the knowledge and experience of the entire team — a valuable source of knowledge and wisdom.
- It is user-friendly and requires no special math!

Kaoru Ishikawa is one of the pioneers of quality management processes. He was at the Kawasaki shipyards, where he became one of the founding fathers of modern management.

The Shape of Cause-and-Effect Diagrams

- A Fishbone Diagram is a methodical and graphical way to identify the root cause of a problem. It is a way to analyze symptoms, organize them and reach a logical conclusion about the cause. It is a way to plot cause and effect. The term "fishbone" is in reference to the appearance of these cause-and-effect diagrams because when completed they resemble the skeleton of fish, with the "head" being the problem statement, and the "bones" being the possible causes of the problem.
- The number of bones can vary from as few as four to as many as six, depending on the complexity of the issue or the major possible areas of the problem. Each major bone is a category of causes; for example, tools.
- Each major bone will generate sub-branches, driven by the thread of the conversation. For example, "the press does not work" may lead to causes such as "the on-off button is broken," which leads to "the contacts are burned out," which leads to "there was a power surge that burned out the contact." As you can see, there are many layers or levels depending of the depth of characterization.
- The number of layers or levels of each major bone depends on reaching the root cause. That is, when you have reached the point where you can diagnose the root cause of the problem, then you need no more layers.

Critical Tools

The Fishbone Diagram

General Rules for Fishbone Diagrams

1. A useful Fishbone Diagram takes two to three hours to complete. Ensure that all participants have the time available. You should not rush it, as it relies on participation and brainstorming.

2. Encourage openness within your group. In some organizations, fear plays a critical role in stifling valuable input. You should reassure the participants that no one will be punished. This is called creating a safe environment. Your own behavior will have a significant impact in the degree of openness.

3. Begin by clearly stating the issue in the form of a problem. That is, you are starting with the effect and searching for the cause. For example, "The manufacturing line stopped at noon." Another example: "The phone rings too many times, and customers are upset."

4. "Why" is your best friend. Constantly asking, "Why?" will encourage the participants to better identify the causes.

The Steps

1. Always start with the effect in the form of a problem statement. For example, "There are too many rejects on station seven" or "The number of customer complaints has risen by 50 percent."

2. Encourage brainstorming to identify the cause of the problem. Reassure people that no one will be punished; it is safe to make suggestions, even if they are not 100 person certain. Encourage off-the-wall statements, and even inject humor and laughter. You must create a safe environment to enable all the information to flow freely.

3. Start by writing the input from the team members on a flip chart or marking board. Once you have enough ideas, start building the fishbone by categorizing the causes into clusters. Generally, there are four primary areas or clusters, although you may choose to have more. We discourage using more than six, as it becomes too difficult to manage. The four most common areas are equipment, people, processes and methods. For manufacturing situations, you might choose people, methods, materials and equipment. For service or business processes, you might choose equipment, policies, processes and people. Or you may replace those terms with ones that are more attuned with your culture and the nature of the problem.

4. Once you finish the categories and have exhausted all ideas, start prioritizing using your own criteria.

5. For complicated issues, you might want to select the top three or four, and have the team conduct additional research. This is particularly applicable in manufacturing or highly technical areas where validation is required.

Important The criteria for prioritizing can be developed on the spot. For example, you might decide to use "1, 2, 3" to represent "most likely source of the problem," where 1 is the most likely. The criteria must be short and easy to understand.

Ask the team to help you prioritize using the newly developed criteria Assign responsibilities once all the prioritization is finished.

General Recommendations for Fishbone Diagrams

- Select the proper team — those people who have visibility of the problem.

- The phrasing of the problem statement is critical, as it sets the tone for the brainstorming. Carefully wordsmith the problem such that it clearly states the real issue. For example, "The press does not work" may not be sufficient. A better way might be to add more detail: "The press in station seven does not drill properly." We suggest that you work with your team to craft a statement that truly focuses on the real issue.

Product Development

The Fishbone Diagram

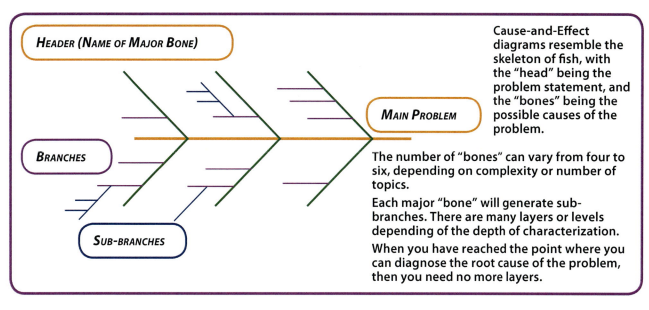

Cause-and-Effect diagrams resemble the skeleton of fish, with the "head" being the problem statement, and the "bones" being the possible causes of the problem.

The number of "bones" can vary from four to six, depending on complexity or number of topics.

Each major "bone" will generate sub-branches. There are many layers or levels depending of the depth of characterization.

When you have reached the point where you can diagnose the root cause of the problem, then you need no more layers.

Detailed Process: Fishbone Cause and Effect

① INVITE PARTICIPANTS

Who should attend?
- All stakeholders in the process or project. Try to keep the group to 15 participants or fewer.
- The ideal number is about 10 participants.

② PREPARE

Checklist:
- A large conference room.
- Post-It Notes® (3M brand, model #655 3x5).
- Sharpie® (Sanford brand, fine point, black markers) for all attendees.
- Flip chart paper and two easels.
- Masking tape.

③ SET UP THE TOPIC

- State the topic in the form of a problem or an issue. For example: "The technical support process does not work." Another example: "The scope of the new software application is not complete."
- Ensure that everyone understands the problem statement. You may need to rephrase it to ensure full understanding.
- Reassure everyone that there are no wrong answers.
- Reassure everyone that grammar and spelling are not important.
- Explain how to use the yellow stickies: one per subject, use the felt-tip pen, landscape orientation and NO criticism.

④ START THE PROCESS

- Given the question or topic, ask everyone to write at least three yellow stickies outlining the reason the problem is occurring. For example, "Why did the technical support process break?" "It broke because there is no automatic answering software."
- When everyone is ready, ask each person to loudly and clearly read their yellow stickies.
- Tell everyone to listen intently as others read their yellow stickies.
- Encourage the participants to write additional yellow stickies as others present theirs.
- Collect all yellow stickies and randomly place them on the flip chart or board.
- Continue until the group runs out of ideas or repetition sets in.

CONTINUED ▶

77

For comments, questions or additional copies please e-mail the author: debriefs@rapidinnovation.com.

Critical Tools

The Fishbone Diagram

CONT'D

5

START THE FISHBONE
- Give instructions to the group.
- Rally the group around the flip charts with the yellow stickies. The group may require encouragement to move up to the board.
- Draw the fishbone; that is, a line with four or six "ribs" emanating from it.
- Ask the participants to help you sort or cluster the yellow stickies into several topics. These topics represent each of the ribs.

6

CREATE THE CATEGORIES (RIB LABELS)
- After you have completed the sorting, label each of the ribs with a header. The header is a name that best represents why the yellow stickies in that particular rib are together.
- This is also a good time to scrub your information. For example, you can delete duplicate yellow stickies and you can rephrase or rewrite some yellow stickies to make them easier to understand.
- Important: some yellow stickies, will actually be a subset or very complementary to others. This represents a branch. So you will have the main ribs, and within each rib you may have branches. You guidance is to layer down to the root cause of the problem.

7

PRIORITIZE
- It is best to prioritize the ribs (headers) rather than each sticky.
- Save the flip charts for archival purposes or to transcribe if it is important to do so.
- Debrief the group on the effectiveness of the tool, and ways to improve the affinity process.

See Also: **Developing Scales for Prioritization**

8

DEVELOP ACTION PLAN
- Assign responsibility for additional work; for example, more research may be needed to validate your findings.
- Develop a plan to ensure that your findings translate to action that will resolve the problem.
- It may be necessary to have additional meetings to continue the process.
- You may need to form sub-teams to address specific areas.
- Above all, leave no loose ends. Ensure closure!

Fishbone Diagram

(diagram showing fishbone with RIB and CATEGORY labels)

There should be from four to six ribs — don't go over six or it becomes too difficult to mange.

Example categories (rib labels):
- People
- Processes
- Methods
- Tools
- Materials
- Equipment
- Information
- Policies

Product Development

The Fishbone Diagram

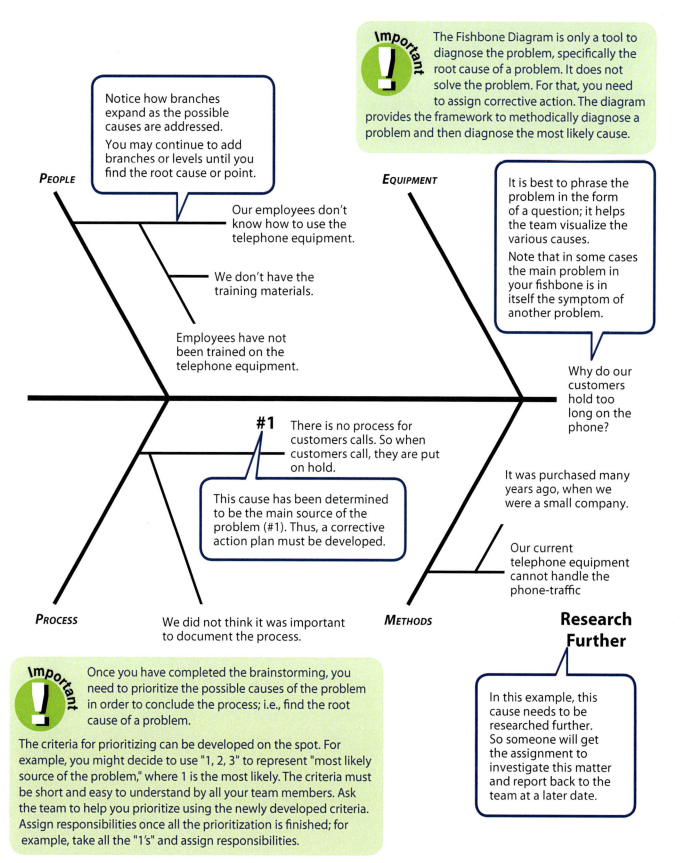

79

The Fishbone Diagram

Developing Scales for Prioritization

Recommendations and Rules

- We highly recommend that you use simple, linear scales from 1 to 10 in every instance. This will keep the focus not on the scales, but the substance of your risk management efforts.

- Always use 1 to indicate the lowest or minimal level, while 10 will always indicate the highest or maximum level. For example, maximum damage or impact of a risk is always 10. Correspondingly, 1 will always indicate the lowest impact. It follows that the probability of a risk occurring will also be highest at a 10 and lowest at a 1. This approach will allow you to obtain information in a harmonized, or aligned, way.

- Refrain from using scales that are too complicated, such as formulas, weighted averages or algorithms. Keep in mind that this is a **QUALITATIVE** effort and adding more resolution will not add more accuracy; on the contrary, it may confuse your team.

- You and your team need to define each step in your scale — for example, 10 means Highly Likely, 8 means Likely, 6 means Moderate, 4 means Occasionally, 2 means Remote and 1 means Unlikely. You must do this in order to allow everyone in your team to have the same perception of a level. Experience will help you decide the amount of definition detail that each level requires.

- We suggest that you start by developing the description of the levels using the even numbers in your scale. This will save you time. As you become more experienced, you can complete the definitions from 1 to 10

- Be certain to document your scales and the definition of the levels. This will allow for repeatability, reference and continuous improvement.

Example: Simple Scales and Definition Levels

A hypothetical scale for probability might describe the levels as shown in the figure below:

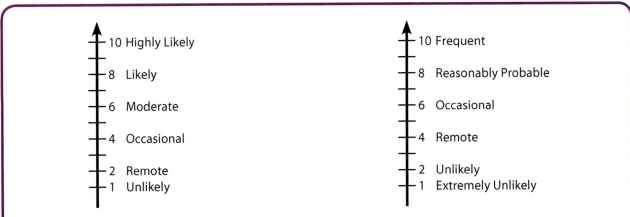

Here are two very simple definitions of a scale for the probability a risk might occur. The meaning of each level (step) of the scales would have been determined by the consensus of the team. You and your team could use either scale as a starting point.

Over time, you can add additional definitions and descriptions among the different levels. Your team will become familiar with the meaning of each number, and the uncertainty between levels will be reduced.

Product Development

The Fishbone Diagram

Example: Sophisticated Scale and Definition Levels

A hypothetical scale for impact or damage might describe the levels as shown in the table below: Over time, you can develop scales that are much more meaningful to your situation.

Level	Rank	Severity of Failure as Seen by the Customer
1	None	No noticeable defects.
2	Very minor	Abnormal noises, noticeable only by the most discerning operators.
3	Minor	Abnormal noises, noticeable by 50% of the customers. Fully operational, but causes stress on the operator.
4	Very low	Abnormal noises, noticeable by the customer. Fully operational, but causes stress on the operator.
5	Low	Operable but difficult to operate, all functions available.
6	Moderate	Operable but difficult to operate, some functions not available.
7	High	Operable, but at reduced level of performance.
8	Very high	Device inoperable.
9	Hazardous with warning	Unsafe to operate.
10	Hazardous without warning	Operator injury very likely.

Here is an example of a more sophisticated definition of a scale, one that focuses on the levels of impact for each risk. The meaning of each level (step) of the scales would have been determined by the consensus of the team.

Critical Tools

Process Mapping

NEW TOPIC

Steps to Success

A process is a **DOCUMENTED**, sequential list of **VALUE-ADDED** steps, the aggregate of which is used to deliver a product to a **CUSTOMER**.

- **DOCUMENTED:** A process must be documented; otherwise, it fails the most critical test. Furthermore, documenting the process has several advantages. It forces the dialog to be collaborative. It allows more than one person to study and use the process. Also, it's the only way to initiate formal improvements.

- **VALUE-ADDED:** Every step in the process must add value, as defined by the customer, or it is waste that must be removed.

- **CUSTOMER:** Every process must have a customer. For example, a customer of Marketing is Engineering. Thus, Marketing is the Process Owner, responsible for creating the product requirements, and Engineering is the Process Customer. Engineering has a set of needs, and it is the responsibility of Marketing to identify and satisfy these needs.

Create a Process Map

Creating of a process is not difficult. You can start by reviewing the many process maps shown in this guidebook. Collaborate with your peers, and define a process using the Yellow Sticky Protocol.

1. Call a meeting with people closely involved in the process you have chosen to document. For the process of producing a new brochure, for example, you would likely invite marketing staff, your advertising manager (marcom) and your advertising agency if you have one. These meetings generally last from two to three hours, sometimes longer.

2. Start the meeting by defining the process you wish to document — in this case, the creation of a new brochure. Also define the start and end of the process. Your team will help you.

3. Define the customer. In our example, the process customer happens to be your company's external customers. Be careful to differentiate between customer and stakeholder. In our example, the marketing manager is not a customer, but a stakeholder with a vested interest in making this process work. Furthermore, your general manager is also a stakeholder.

4. Define the output (result or product) of using the process. In our example, it may be "a printed brochure, ready for delivery to our customers." This critical step clearly defines the endpoint.

5. Brainstorm with your team to identify all the steps. Using yellow stickies, one for every step. At this point, you don't need to be concerned with value, sequence, duration, etc. of the steps.

6. Organize the steps sequentially, from process start to end. Add and refine steps as needed. Eliminate redundant or trivial steps. Take your time; this adds much clarity to your process map.

7. Study your map to ensure accuracy and clarity. Is the sequence of steps the way things really happen? Are there any steps that don't make sense? Are the steps described in a way that's easy to understand?

8. Test every step against the definition of the process. Pay particular attention to the value-added criteria. Ask yourself, is this step necessary? Does it really add value? This is your first opportunity to improve efficiency.

9. Finally, review your process map one last time, document it for permanent use and start using it.

Continuous Improvement

Once you begin using your documented process, the payoff begins. But the payoff continues as you make continues improvements to your process map. As you use it you will discover obvious ways to improve it — for example, redefine, reorder or eliminate steps. The secret is to constantly review your process map to identify opportunities for improvement.

Appendixes

Appendix A: Recommended Books 84
These lists reference books that the senior consultants at RapidInnovation, LLC have read and found applicable

Appendix B: Glossary 91
Terms you will encounter in the real world of product development explained in a practical, common-sense fashion. With this resource, you will be able to better connect the concepts and tools of product development to your projects.

Appendix C: Templates 114

Recommended Books

Introduction

These lists reference books that the senior consultants at Rapidinnovation, LLC have read and found applicable to today's business challenges. Rapidinnovation, LLC is a consulting firm whose primary focus is serving product development organizations around the world, which we've been doing since 1997, and with many satisfied clients.

We welcome your recommendations and reviews of these and other books and resources. Please e-mail the authors with your comments and suggestions.

Teams and Teamwork for Product Development

Book Title & Author	Subject	Comments
The Wisdom of Teams By Jon R. Katzenback Published by Harvard Business School Press	• High-performance teams.	• Excellent book for creating development teams. • Clear, no-nonsense approach to cross-functional teams such as concurrent engineering, product development, etc.
Peopleware, Productive Projects and Teams, Second Ed. By Tom DeMarco & Timothy Lister Published by Dorset House Publishing	• Software development. • People management.	• Good book to help understand the human side of software development.
Principle-Centered Leadership By Stephen R. Covey Published by Fireside	• Timeless principles of leadership. • Clear, no-nonsense understanding of leadership in a business environment.	• The difference between a manager and a leader, and its implications in an organization.
Punished by Rewards By Alfie Kohn Published by Houghton Mifflin Company	• Reward, recognition, celebrations and motivation in organizations.	• The truth about failed recognition programs and the unintended results of poor motivation and recognition programs.
Mastering Virtual Teams, Third Ed. By Debora L. Duarte and Nancy Tennant Published by Jossey-Bass	• The ways and means for enabling virtual teams to succeed.	• Duarte does a good job of presenting the challenges and solutions for distributed teams. A level-headed presentation of the issue.

Appendix A

Recommended Books

General Management and Strategy

Book Title & Author	Subject	Comments
The Innovator's Dilemma By Clayton M. Christensen Published by HarperBusiness	• Why companies fail to stay at the top of their industries.	• A book that confronts a reality faced by most successful technology companies: the need to change, and change fast.
The Discipline of Market Leaders By Michael Treacy Published by Perseus Books	• Choose your customers, narrow your focus, dominate your market.	• Deliberately choosing the customers you wish to serve allows you to develop the competencies needed to succeed.
Built to Last By James C. Collins Published by HarperBusiness	• Visionary companies. • Values, culture and successful businesses.	• Focusses on the human side of corporate culture for the promotion of long-term success.
The Balanced Score Card By Robert S. Kaplan and David P. Norton Published by Harvard Business Press	• What to measure and why. • Financial measures alone are not enough to ensure sustainable customer satisfaction and profits.	• Rational approach to measures. • Systemic look at the enterprise. • Focus on the human side of measures.
Companies with a Conscience, Third Ed. By Mary Scott and Howard Rothman Published by Citadel Press	• Case studies of 12 companies that are profitable and at the same time people-, environment- and community-friendly	• Good examples of "having your cake and it eating it too" — the ability to make a profit and have a clear corporate conscience.

Product Development

Book Title & Author	Subject	Comments
Rapid Development: Taming Wild Software Schedules By Steve McConnell Published by Microsoft Press	• Perspectives and methods for rapid software development. • Focus on the process and framework for development.	• The methodical approach to rapid software development. • Good food for thought.
Revolutionizing Product Development By Steven C. Wheelwright Published by The Free Press	• One of the classics on product development.	• A must-have for those who are serious about product development.
The Invisible Computer By Donald A. Norman Published by The MIT Press	• Why good products can fail.	• Insightful as to why technology companies behave like technology companies.
Dynamics of Software Development By Jim McCarthy Published by Microsoft Press	• Rapid software development is not the result of countless hours of coding.	• A book more about teams and teamwork than software development.

Recommended Books

Book Title & Author	Subject	Comments
Portfolio Management for New Products, Second Ed. By Robert G. Cooper, Scott J. Edgett and Elko J. Kleinschmidt Published by Perseus Books	• Managing of technology for profits. • Disciplined product development. • Separate the losers from the winners.	• A solid, direct book on managing a portfolio of products while optimizing resources. • Good methods for balancing your product offering.

Productivity Improvement

Book Title & Author	Subject	Comments
The Fifth Discipline By Peter M. Senge Published by Doubleday	• System thinking. • The quintessential book of systemic approaches to solve business problems.	• A must-read to understand the total solution to customers' problems • Also very important to re-engineering
Improving Performance: How to Manage the White Space on the Organization Chart By Geary A. Rummler Published by Jossey-Bass	• Business processes. • Teamwork in organizations. • Process mapping for non-manufacturing organizations.	• Clear understanding of processes in a business environment. • Good how-to information and a level-headed approach to organizational development.
Kaizen: The Key to Japan's Competitive Success By Masaaki Imai Published by McGraw Hill	• Continuous improvement.	• The truth about continuous improvement as a culture and not a fad.

Appendix A

Recommended Books

Project and Program Management

Book Title & Author	Subject	Comments
Flexible Project Management for Product Development By Jose Campos Published by Rapidinnovation, LLC	• Clear understanding of the conventional project management rules and tools do not necessarily work in a product development environment. But here, the author has experience in both project management and product development.	• Easy to understand and very detailed. • Plenty of examples, templates and forms ready to use. • Easy to share with other members of the development team. • Clearly written with product development in mind, very practical.
Project Management Toolbox By Dragan Milosevic Published by John Wiley and Sons	• A comprehensive set of project management tools full explanations and plenty of examples	• A must-have reference
Program Management for Improved Business Results By Dragan Milosevic, Russ Martinelli and James Waddell Published by John Wiley and Sons	• A business approach to the role of program managers. Authored by Dragan Milosevic, acclaimed authority in project and program management.	• Probably the best book on program management.
Reinventing Project Management By: Aaron J. Shenhar Published by Harvard Business Press	• Proof positive that every project is different and therefore, you cannot manage them exactly the same.	
The Fast Forward MBA in Project Management, 2008 Ed. By Eric Verzuh Published by John Wiley and Sons	• Looks at the challenges of today's project environment and examines the skills a project leader needs to make a project successful. • It includes organizational changes, tools and techniques, control methods and real-life applications.	• Good medium-depth overview of all aspects of project management. • Easy to read, with examples and case studies, but meaty enough that both junior and experienced project leaders will find nuggets. • Down-to-earth
The New Project Management, Second Ed. By J. Davidson Frame Published by Jossey-Bass	• Frame makes a case that conventional wisdom is only marginally relevant to project leaders working in today's turbulent project environments. • He offers strategies and techniques for changing from implementor to initiator, and for dealing with the soft and complex issues that affect most technology projects.	• Good ideas, especially on soft skills and orientation toward the customer. • Frame sometimes undervalues the traditional techniques and gets too enthusiastic about how empowered a project leader can be. Pair this book with a solid grounding in the hard fundamentals of project management, such as from Verzuh's book.

Recommended Books

Book Title & Author	Subject	Comments
Visualizing Project Management, Third Ed. By Forssberg, Mooz and Cotterman Published by John Wiley and Sons	• A recently updated project management textbook with lots of examples. It strives to provide an integrated model of project management, and focuses on vocabulary, teamwork, the project cycle and management techniques.	• Covers a wide variety of hard and soft project management topics, with short case studies of how they have been used in real situations. • Focuses mostly on applying them within a sequential framework, so don't look for lots of insight on iterative approaches here. The authors' experience in aerospace and government projects shows through in the vocabulary and examples.
The Wiley Guide to Managing Projects By Peter W. G. Morris and Jeffrey K. Pinto Published by John Wiley and Sons	• Published in 2005, this book is up-dated every 10 years. This is a comprehensive book on project management. Looks well beyond the basics on project management (cost, schedule and performance).	• A must-have reference for those who are serious about project management. • Critically important for those involved in the strategies of project management.

The Voice of the Customer (VOC)

Book Title & Author	Subject	Comments
The Voice of the Customer for Product Development By Jose Campos, Jean-Claude Balland Ph.D. Published by Rapidinnovation, LLC	• A detailed guidebook on how to capture and use the voice of the customers (VOC) for product development. The author describes every step and detail in an easy to understand fashion.	• Easy to understand and very detailed. • Plenty of examples, templates and forms ready to use. • Easy to share with other members of the development team.
How Customers Think By Gerald Zaltman Published by Harvard Business School	• Essential insight into the mind of the customers and the market.	• Customer needs are much more subjective than we thought. • An eye opener concerning value.
Advanced QFD By Larry Shillito Published by Wiley & Sons, Inc.	• The use of QFD (Quality Function Deployment) in a real-world environment, linking technology to the market.	• A sound, levelheaded book on QFD, A must-have if you are serious about QFD.
Software Requirements, Second Ed. and More About Software Requirements By Karl Wiegers, Published by Microsoft Press	• A practical guide for gathering and managing requirements throughout the product development cycle.	• As relevant to product requirements as it is to software requirements, this practical guide will go a long way towards making your R&D and Marketing people see the world through the same lens.

Appendix A

Recommended Books

Book Title & Author	Subject	Comments
Customer-centric Product Definition By Sheila Mello Published by AMACOM	• Good coverage of the Voice of the Customer (VOC) and its use in defining new products.	• Very good book; the author clearly portrays the process of getting the customers at the center of new product development.
Writing Effective Use Cases By Alistair Cockburn Published by Addison Wesley	• A practical guide to the powerful approach of expressing the behavioral requirements of complex systems.	• Practical, with numerous examples of good and bad use cases.
Customer Visits, Second Ed. By Edward F. McQuarrie Published by Sage Publications	• A clear description of what it takes to perform a good customer visit.	• The author is well-recognized as an authority in this field. Levelheaded approach.
Acquiring, Processing, and Deploying the Voice of the Customer By M. Larry Shillito Published by St. Lucie Press	• Description of ways to obtain, process and use the voice of the customer.	• Plain English, down-to-earth, but clearly from someone who knows.
Voices Into Choices By Gary Burchill Published by Joiner	• A handbook with step-by-step descriptions to obtain and use the voice of the customer	• Comprehensive. • Truly a fine work — a must-have handbook.
Priceless: Turning Ordinary Products Into Extraordinary Experience By D. LaSalle & T. A. Britton Published by Harvard Business Press, 2003	• The title says it all. People don't buy products, they buy value and experience. This book will help you see the world through the eyes and heart of customers.	• Very valuable in bringing the new concept of experience to a tactical level.
Quality Function Deployment: How to Make QFD Work for You By Lou Cohen Published by Addison-Wesley, 1995	• A tour of QFD fundamentals, as well as advanced concepts	• Clear exposé to a sophisticated subject. • Most useful to engineers than marketers.
The Observational Research Handbook: Understanding How Consumers Live with Your Product By Bill Abrams Published by McGraw-Hill Professional, 2000	• Practical coverage of the art of observing customers in context to understanding latent needs.	• Useful advice on how and when to observe customers to get clues to their needs.
Learning from Strangers: The Art and Method of Qualitative Interview Studies By Robert S. Weiss Published by Simon & Schuster, 1995	• The subtitle of the book says it all: "The art and method of qualitative interview studies." These are important skills to master.	• A how-to book on interviewing customers by an expert on the topic.

Recommended Books

Risk Management and FMEA (FMCA)

Book Title & Author	Subject	Comments
Risk Management and FMEA for Product Development By: Jose Campos Published by Rapidinnovation. LLC	• A detailed guidebook on risk management for product development. The author describes every step and detail in an easy to understand fashion.	• Easy to understand and very detailed. • Plenty of examples, templates and forms ready to use.
Waltzing with Bears: Managing Risk on Software Projects By Tom DeMarco and Timothy Lister Published by Dorset House Publishing		
Proactive Risk Management: Controlling Uncertainty in Product Development By Preston G. Smith and Guy M. Merritt Published by Productivity Press	• A "plain English" book on risk management.	• A good reference for anyone responsible for managing risk.
Managing the Unknown By Christoph H. Lock, Arnound DeMeyer and Michael T. Pich Published by John Wiley & Son	• A book about managing the risks you don't know: the ways and means to deal with total uncertainty. • Intriguing new concepts about risk, particularly in areas with high uncertainty such as new technology.	• A good book to have if you are in product development or technology development. • Challenges conventional views of risk management.
Project and Program Risk Management: A Guide to Managing Project Risks and Opportunities By R. Max Widerman Published by Project Management Institute, 1992	• A risk management book for beginners and intermediate managers of risk.	
Failure Mode Effect Analysis: FMEA from Theory to Execution, Second Ed. By D. H. Stamatis Published by American Society for Quality, 2003		• A comprehensive book on FMEA, strong in context and advanced methods. • A good overall reference book for those serious about FMEA.
Project Management Tool Box By Dragan Milosevic Published by John Wiley and Sons, 2003	• A comprehensive set of project management tools, full explanations and plenty of examples.	• A must-have reference.

Appendix B

Glossary

Definitions for Product Development

Practical Definitions in Plain English to Enable Effective Action

The intent of this glossary is to explain the terms you will encounter in the real world of product development. Furthermore, we explain these terms in a practical, common-sense fashion. When multiple relevant definitions exist, we will provide them. With this resource, you will be able to better connect the concepts and tools of product development to your projects.

Term	Definition & Notes
Agile	• A software development model or framework based on frequent iterations with the customer and development of the software in small, incremental amounts, collaboration and flexibility.
Alpha Unit	• A version of a new product that is ready for internal testing. The product meets criteria that enable testing that will yield valuable data and input to proceed with the final version. • After the Alpha Unit comes the Beta Unit, a more mature version of the product. • In hardware an alpha unit may be used to test the manufacturing process, functionality or testing strategy. • In software it can be used to test human interface (GUI), usability or functionality. • Alpha units are not saleable and are intended for internal use only.

Assumption(s)	• A short set of relevant factors that can have a positive or negative impact on your project or program and define the playing field.
	   – The playing field refers to the arena or area of focus for your endeavor.
	   – A short set means three to seven relevant factors.
	   – Relevant means of importance to you and your team — that which must be taken into consideration when making decisions.

Attribute	• Used to define a set of general characteristics of a platform, product line or product, an aggregate of several features. Examples include "ease of use," "ruggedness" and "re-configurable". • Related to, but not the same as "Product feature." See also "Feature."
Awareness	• A measure of the percent of target customers who are aware of the existence of something. It could be a product, a brand or even a concept. • Awareness does not imply preference. • Awareness can be measured for brand, product or any other attribute you select. • Awareness should be measured before and after an event; for example, an advertising campaign. The difference provides a measure of effectiveness. • There are two kinds of awareness: Unaided awareness is when the customer volunteers the information (manifests the awareness) without any help. Aided awareness is when a hint or suggestion is provided, such as a photograph, and then the customer expresses awareness.

Glossary

Term	Definition & Notes
Benchmarking	• The process of studying how successful companies, industries and even competitors perform specific functions and then selecting the best as standards of performance for your own improvements. • You can benchmark with your competitors or across industries, in your own country or across the world. • Benchmarking is a good way to develop your level of excellence, and to learn "good practices" for your continuous improvement.
Best Practices	• A method, technique, tool or other practice that has shown to be the best within a certain category. Best practices must be researched, defensible and published by an authorized organization. • Methods, approaches or processes deemed to be "best in class" and, consequently, worthy of imitation. • See also "Good Practices."
Beta Test (a.k.a. Beta Unit or Beta Version)	• An external test of pre-production products. A way to test new products in actual application to discover bugs and last minute fixes before full introduction. • Some companies (software in particular) use Beta Tests as part of their introduction strategy, to generate awareness and word of mouth.
Bill of Material (BOM)	• A hierarchical list of systems, sub systems, sub assemblies and components that make up a product or a higher level sub-system. • It can be a set of parts list, drawings, and schematics. In manufacturing it can also mean the way a new product is assembled including all parts, systems and sub systems.
Brainstorming	• The process of tapping into a group's creativity to generate ideas, solutions or innovation. • Generally, it takes the form of a meeting or gathering where the team or group is encouraged to come up with as many creative ideas as possible. • There are many methods to facilitate or enable brainstorming. The methods are intended to enable the ideation, processing of the information, and finally prioritization of the ideas. Many brainstorming sessions fail due to a lack of a framework or process.
Brand	• The combination of all tangibles, intangibles and actions of a company that create a particular promise of value in the customers' mind. • A brand is a holistic concept that evokes the mental images and expectations the customers have about a particular company.
Breakeven	• A financial metric that identifies the point where the revenue (sales) of a new product match the investment made in developing it. • At this point, the new product has reached enough sales to start generating "positive cash flow."

Appendix B

Glossary

Term	Definition & Notes
Business Case	• The results of the business analysis to demonstrate the viability of a new product or service prior to approval for development. • Viability includes profitability, market share or other critical and strategic objectives of your company. • "Making the business case" in product development implies your ability to convince the company executives that investing in your new product will result in a measurable return. Philosophically, you are going to the company executives to ask for an investment, which will results is a quantifiable return
Business Model	• The ways and means selected by the company to generate revenue and make a profit. • The Business Model is said to be working when the company has deployed and executed the strategies to generate revenue, profits and serve its customers.
Business to Business (B-to-B)	• Sales to other businesses and not to individual consumers. For example, an OEM sells components to a manufacturing company. • Sales to distributors, representatives or wholesalers are also considered B-to-B.
Champion	• A person who takes strong personal interest in seeing that a particular process, product or other is carried to fruition. Champions are always ready to deal with adversity and committed to success. • Champions are said to be "maniacal" in their commitment, willing to put exceptional effort and energy into their cause. • No innovation can be brought to fruition without champions.
Change Management	• An organizational process aimed at helping employees embrace changes in their current business environment and then change their behavior in support of the new state. • A structured approach to shifting or transitioning individuals, teams and organizations from a current state of behavior to a desired future state.
Charter, Team Charter	• The document that officially starts an initiative. This document outlines the team members, the common purpose that brings them together and the performance goals to ensure progress and outcomes. It also defines the roles, responsibilities and authority. It defines the desired approach to ensure completion. • The team members should be strongly involved in — even responsible for — the development of the charter. • The Charter must be fully supported by Management. • A charter is the unifying principle of the team. Without it, each team member has a different idea of what needs to get done and how.
Co-location	• Product development where the entire development team is in one location; for example, the same building. • The processes and methods for co-location development are different from that used for multi-site or distributed product development.
Commercialization	• The process of taking a new product to market. The transformation of an idea into a product that customers are willing to purchase. • Commercialization can be used for a new product, a new technology or anything that a particular set of customers value and are willing to pay to obtain it.

Glossary

Term	Definition & Notes
Compelling Reason To Act (CRTA)	• The needs the customer must address now so that a budget and the will to act can be generated. • CRTA is proposed to designate the highest priority needs — the ones that can be funded and will be acted upon by your customer. • A customer must have a CRTA in order to purchase your product. It is your responsibility to identify CRTAs so you can develop products that address customer requirements.
Competency	• The aggregate set of skills, knowledge and abilities that an individual needs to fulfill a specific set of measurable responsibilities. • Competency is all about one individual — not a team or an organization. The equivalent statement at the organizational or team level is known as "maturity" or "organizational maturity"
Complete Product	• The aggregate of the tangible, intangible and emotional components that constitute a product in the eyes of the customer. • See also "Offering" and "Whole Product."
Completion	• The verifiable fulfillment of all objectives. • The conclusion of all efforts; for example, the completion of a new product development is the introduction or rollout to the market place.
Compound Annual Growth Rate (CARG)	• The calculation of rate at which an investment would grow if it grew at a steady rate. It's a way to smooth the return on an investment for the purposes of making informed decisions. • In product development, CARG is generally used to describe the growth rate of a market or the growth rate of new products.
Computer Aided Design (CAD)	• The use of computers and software to design a new product or parts of a new product. • CAD is a very advanced technology enabling simulation, testing, and all aspects of design.
Computer Aided Manufacturing (CAM)	• The use of computers and software to design and run a manufacturing process in an integrated fashion; that is, as a system. • In some areas CAM is also used to describe any aspect of the manufacturing process that is run with the aid of computers or fully automatic, not necessarily the entire system of manufacturing. • Widely used in industry, CAM has become essential to running manufacturing operations around the world. • CIM (Computer Integrated Manufacturing) may be used in conjunction with CAM.
Computer Integrated Manufacturing (CIM)	• See "Computer Aided Manufacturing (CAM)"
Concurrency	• The ability to develop several products at the same time. • The goal of a product company is to achieve speed, concurrency and repeatability. • It can also be applied to the development of a single product where parts, assemblies and subassemblies are developed at the same time and not sequentially.

Appendix B

Glossary

Term	Definition & Notes
Concurrent Engineering	• A model of product development where all functions such as engineering, marketing, manufacturing, etc., work at the same time to design and take the product to market. • A set of methodologies that enable rapid product development by allowing as many activities as possible to take place simultaneously. • Contrasted to the sequential model, where one activity had to wait for another before it could be started.
Concurrent Engineering	• An emerging engineering methodology based on the premises: – All aspects of the life-cycle of a product should be considered as part of the initial design process; that is, from basic functionality until end of life. – All activities that can happen at the same time must be undertaken. • The methodology, if properly applied, yields better products in a shorter amount of time.
Conjoint Analysis	• A quantitative market research technique used to characterize how people value the features (or benefits) of a product. • One of the few methods available to determine the possible price of a new product with objectivity. • Conjoint analysis is powerful because it puts the customer in a real life situation of having to make trade-offs between benefits (what he wants to get) and costs (what he has to give to get it).
Constraint	• Restriction or boundary imposed on a project or design. • Constraints may help in setting priorities, avoiding risks or the physical limitation of the technology.
Contextual Inquiry	• A way to gain knowledge about a customer by observing people in their environment doing specific activities. • The use of anthropology, social science and other techniques to study customers performing specific tasks for the purpose of understanding their needs. • The observation must be qualitative.
Core Competency	• The few things that a company does that: – Its customers find exceptionally valuable. – It does better than its competitors. – Its competitors find difficult to imitate. – It would never outsource to anyone. – It can transfer to other industries as is. – Is part of its growth plan. • The definition of Core Competency varies. It's driven by the specific situation of a company at the time it defines its Core Competencies. Nonetheless, it is critical for a company to understand and develop its core competencies. • Core competencies are always a web of things that a company does particularly well without sometimes even be conscious of it.

Glossary

Term	Definition & Notes
Core Team: Concurrent Engineering	• A small group of representatives from the critical functions (e.g., Engineering, Marketing, Program Management, Finance, etc.) responsible for the development of a specific product and accountable to deliver measurable results. • The Core Team is responsible and accountable for progress and eventual delivery of the new product to the market. It acts as the Program Manger for the development.
Core Team: Generic	• A small group of three to five individuals that is chartered and empowered to guide the implementation and completion of an initiative. • Very useful in fast-changing environments where distractions abound because a small group can focus attention and energy to ensure clean execution.
Corrective Action	• The specific steps to eliminate the cause of a detected nonconformity or other undesirable situation. That it, the steps that are taken to permanently remove the causes of an existing nonconformity through root cause analysis and deploying permanent corrections. • In contrast, "Remedial Action" refers to the temporary steps taken to eliminate one specific non-conformance without regards to a permanent solution.
Criteria	• A standard used to evaluate alternatives in a decision. Criteria are always finite and easy to understand by all. • In the product development process, the criteria are generally defined by the Gatekeeper at major gates.
Critical Path (CP)	• The path with the longest duration • CP comes from the analysis of a network diagram or a PERT (Project Evaluation and Review Technique) diagram in project management.
Critical Success Factor (CSF)	• Something that is critical to your program and must happen. • The few key areas where things must go right for the business to flourish. If results in these areas are not adequate, the organization's efforts for the period will be less than desired. • Those few things that must happen in order to achieve the desired results.
Cross Section (Market)	• A typical selection or sample. For example, a cross section of a particular market segment. • This term is a metaphor from engineering, used to reflect the framework of a particular structure such as a market segment.
Cross-Functional Team (a.k.a Multi-Functional Team)	• A team made up of representatives from various functions. For example, marketing, engineering and manufacturing. • Most product development is achieved though cross-functional teams.

Appendix B

Glossary

Term	Definition & Notes

Culture	• A mosaic of interrelated elements and their collective day-to-day interaction that shape the behaviors of people in the group and what's right and wrong.
	• For example:
	  – Management behavior.
	  – Customs, myths and norms.
	  – Ceremonies, rituals and events.
	  – Rewards and consequences.
	  – Physical environment.
	  – Rules and policies.
	• A company culture develops when corporate values are consistently followed for a number of years.
	• Said another way: "Culture Consists Of A Shared, Commonly Held Body Of General Beliefs And Values That Define The 'Shoulds' And The 'Oughts' Of Life." — Kluckhohn

Customer	• The person or entity that pays for the value of a product or service. • There is tendency in some organizations to refer to teams and functional groups as "internal customers." This tends to diffuse the importance of the one and only true customer, the one that pays. • Customers can be "users" or not of your product. Make the distinction between the person who pays and the person who uses.
Customer Experience	• A succession of events in the life of the customer dealing with an offering. This interaction will generate good or bad feelings. • Designing experiences is the new frontier in user-centered design because creating pleasant feelings is the best way to generate satisfaction, loyalty and word of mouth. • Thinking in terms of customer experience forces the marketer to consider the entire sequence of events that the target customer undergoes when dealing with a company and its offerings.
Customer Intimacy	• The business definition as introduced by Michael Treacy in Discipline of Market Leaders is (paraphrase): The discipline of delivering what each specific customer wants or needs rather than what the market as a whole wants or needs. • Customer-intimate companies develop deep relationships with their customers so as to anticipate their needs and solve their problems better and faster.
Customer Requirements	• The crystal-clear articulation of the desired outcomes a customer is striving to achieve.
Customer Satisfaction	• The positive or negative evaluation or feeling of customers that results from comparing the actual experience with a product or service with the expectations it had prior to purchase and use. • Customer satisfaction is a different concept than "customer value." A customer may be satisfied or not with the value it gets.
Customer Value	• The perception of what customers will get (benefits) for what they have to give (costs). • Alternatively, it is the customers' perception of the net benefits they will derive from your offering at a particular price, situation and use.

Glossary

Term	Definition & Notes
Customer Value Drivers (CVDs)	• The criteria that are perceived by the customer to be the most important to the buying process. The yard stick that the customer uses to make the buy decision. • Customer value drivers can be tangible (objective) or intangible (subjective).
Customer-Driven	• The urge to grow a business culture committed to the relentless creation of superior value for the customer. • It is a set of values that put the customers' interest first in order to develop a long-term profitable enterprise — while not excluding those of other stakeholders (paraphrased from: Narver and Slater). • In a for-profit business it is part of the overall strategy to generate revenue and profits. A company can make a profit and not be customer-driven.
Defect	• Lack of conformance to a standard. • Since the introduction of Quality methods, a defect is anything that your customer says it is. This is an effort to focus the organization to evaluate products and services as the customer would.
Defects: Engineering	• Engineering "defects" include design errors, software bugs, incorrect documentation and any substandard design or development steps that would not occur if the process were under control.
Deliverables	• The visible or "touchable" results from a task or a project. The outcomes from a task or a project. • Deliverable has come to mean more than "able to be delivered", now more at "the results of an endeavor or a project." • For example, the deliverable from the marketing organizations is a Marketing Requirements Document (MRD).
Delphi Process	• A technique that uses iterative rounds of consensus development across a group of experts to arrive at the most probable outcome for some future state. • "Experts" are people who have knowledge in a particular area; for example, knowledge about a specific market segment.
Derivative Product	• A new product based on changes to an existing product without affecting the basic product architecture or platform. • A derivative product can be an improvement to an existing product or a different version to satisfy a specific set of customers. • In software development, a rev includes new value to the customer. A rev to fix bugs is not a derivative product.

Design Change Order (DCO)	• A process to ensure the smooth and controlled change of a design and minimizing disruption.
	• Documented changes to the design of a new process, which is part of a formal process for changes.
	• The purpose of the DCOs is to minimize disruptions to the design process and ensure clarity of communication.
	• Also used in this context:
	  – ECN: Engineering Change Notice
	  – ECC: Engineering Change Control
	  – ECO: Engineering Change Order

Appendix B

Glossary

Term	Definition & Notes
Design Engineer	• A person with proven domain expertise and experience who is responsible for creating innovative solutions that customers value. • The output of a design engineer is a package of technology in the form of documentation (drawings) that is the tangible manifestation of an innovative solution that customers value.
Differentiated Product	• Characterizes a product that provides more advantages than its competitors for the most important Customer Value Drivers of a given segment. • A well-differentiated product is one of the most important success factor for a new product.
DOA (Dead on Arrival)	• Products that arrive at a customer site unusable. • A product that does not work when the customer unpacks it.
Early Adopters	• Customers who buy into a new product or technology very early in its life cycle. • Early Adopters are willing to take the risk of an unproven new product or technology to gain a business or personal advantage.
Economic Value Analyses (EVA)	• A methodology that enables you to understand the real progress of a development process by measuring actual progress towards completion. A way to clearly understand progress towards the objectives of a new product. • EVA is found in project management texts — very popular in the construction industry — and now being adopted in product development due to its objectivity. • While budge and schedule are indicators of "movement", they do not necessarily show progress towards the achievement of the objectives of the new product.
Elevator Pitch (a.k.a. 30-Sec. Commercial)	• A short statement summarizing in a compelling way a value proposition. • If a pitch can't interest a prospect in the time it takes to move a few floors in an elevator, you don't have a strong, well-differentiated offering. Rework the pitch and try again.
Engineering Change Control (ECC), Engineering Change Notice (ECN), Engineering Change Order (ECO)	• A formal process that documents a needed design change regardless of the root cause of such change. The process must follow sound engineering principles for control, tracking and disposition. • See "Design Change Order (DOC)." • EISs may include modeling, simulation, analysis and other computational requirements.
Engineering Information System (EIS)	• Applications and solutions for information-intensive engineering and scientific endeavors. The ability to model, process and manage very large amounts of information for the purpose of enabling engineers and scientists to accomplish their goals.
Experience	• A succession of events in the life of the customer dealing with an offering. This interaction will generate good or bad feelings. • Designing experiences is the new frontier in user-centered design because creating pleasant feelings is the best way to generate satisfaction, loyalty and word of mouth. • Thinking in term of customer experience forces the marketer to consider the entire sequence of events that the target customer undergoes when dealing with a company and its offerings.

Glossary

Term	Definition & Notes
Failure	• A defect, nonperformance or lack of adherence to a specific requirement. • Something that causes a product to no longer fulfill its intended function.
Failure Analysis	• A process to methodically characterize and determine the root cause of a failure. The reasoning from the symptoms of a failure to the identification of the root cause of the failure. • Used in reliability engineering, as well as FMEA (Failure Mode and Effects Analysis).
Feature Creep	• The tendency to add new features and attributes to a new product during the design process. • Feature Creep is not a desirable activity as it delays the product, adds to its complexity and may even interfere with the business objectives of the new product. It also increases the stress and aggravation of the entire development team.
Feature (a.k.a Attribute)	• A quality or capability of an offering. • When they are familiar with the product category, customers often express their needs in term of features or attributes (e.g., "I need a 4-door sedan that gets more than 25 mpg") making it more difficult to understand their primary needs.
Firefighting	• The act of addressing too many interruptions, trivial issues and other distraction at the expense of the schedule. • Taking care of surprises and other unexpected distractions. • Firefighting is generally caused by a severe lack of planning and the lack of a risk management plan.
FMEA (Failure Mode and Effects Analysis)	• A structured process to identify potential failures of new products, components or subsystems. A form of risk management. • Can also be used to analyze and determine the root cause of actual failures.
Focus Groups	• A market research tool where a qualified facilitator (moderator) leads a discussion with a select group of target customers for the purpose of understanding their needs and preferences. • Focus groups are generally eight to twelve customers in a single room; generally the room is designed for focus groups. The facilitator (moderator) leads the conversation, and the response is observed and also video taped for later analysis. • Focus groups are a qualitative form of market research.
Functional Test	• Testing a circuit board to identify functional level faults or defects.
Funnel	• See "Roadmap."

Appendix B

Glossary

Term	Definition & Notes
Fuzzy Front End	• The beginning of a program or project when much is not known and a high level of confusion exists. It's also the best opportunity to influence the long term success of the program or project. All programs and projects have this phase. • Initially, this term was defined as the period of time when the development of a new technology was not clear (fuzzy). The meaning has been extended to include the start of a product development project when much confusion exists — more questions than answers. • We suggest that every development project has a fuzzy-front-end; consequently, it is best to acknowledge it and set about exiting this phase by obtaining the missing information
Gantt Chart	• A horizontal bar chart used in project scheduling that shows the start and end dates of tasks within a project. • A Gantt Chart is a very common way to depict project schedules due to its simplicity and visual impact.
Gate	• The decision point, generally a meeting, at which a management decision is made to allow the product development project to proceed to the next gate. • A gate has only three outcomes: "YES", "NO" or "Delay." • Gates are generally related to the timeline of the project; consequently a gate has a chronological impact. • The metaphor of a gate is part of the Gate Phase approach to project management, which has been part of product development process for many years. This approach is now more likely replace by "Agile" and other forms of iterative product development.
Gate Phase	• A product development process or methodology based on specific stages, where at the end of one stage and prior to the start of the subsequent stage there is a review to ensure that a set of criteria has been met.
Gatekeeper	• The person who or organization that holds approval. • In product development, the person who or organization that decides if the product development team can proceed with a new stage if the criteria from the previous stage has been met.
Go/No Go Gate	• A gate in the development process that allows only two outcomes: proceed or stop. • Critical in any development process, as it determines if the development meets the criteria to proceed to the next gate in the process. • A "no go" decision does not necessarily mean end of the development project.
Goal	• A purpose toward which a project is aimed. • A general direction of achievement that is not measurable. Rather, it is statement of intent; for example, "increase sales" or "accelerate product development." • We distinguish between "goal" (immeasurable achievement) and "objective" (measurable achievement). Some choose to reverse these terms.
Good Practices	• Obtained by establishing benchmarks to identify what meets a criteria of excellence. • See also "Best Practices."

Glossary

Term	Definition & Notes
High-Performance Team	• A small number of people with complementary skills who are committed to a common purpose, performance goals and approach for which they hold themselves mutually accountable — and are deeply committed to one another's personal growth and success. — from The Wisdom of Teams, by Katzenback.
Hurdle Rate	• The minimum return on investment or internal rate of return a new product must meet or exceed as it is considered for approval and proceeds through development. • A Hurdle rate can be applied to any investment.
Ideation	• Similar to brainstorming but focused on generating specific ideas about improving the performance of a product or service. • The process of idea generation generally occurs at the early stages of development.
Inbound Marketing	• The collection of all activities that provide information about the market, including market research, customer research, feedback from other organizations like sales, training and support, industry reports, competitive analysis and technology assessments. • The function responsible for obtaining and prioritizing customer requirements, then collaborating with engineering to develop innovative solutions. • Performing both tactical competitive analysis (product-to-product) and strategic competitive analysis (company-to-company).
Infant Mortality	• Failure of a new product at the early stages of use by the customer. • It can be a metric of the rate of early failure of a new product. • Not to be confused with DOA (Dead on Arrival), which is a different metric.
Innovation	• The introduction of new capabilities that create value for customers. In the business world, innovation is so only if the customer says it is — only if the customer sees value in your offering. • The ability to deliver new value to customers across the entire set of customer experiences — from sales, to delivery, to support, operations and beyond. Every "touch point" with customers provides ample opportunity for innovation and thus value creation. • It's one thing to think up an innovation; it's another to develop it and deliver it to the customer. • Innovation is clearly not limited to technology!
Intangibles	• Attributes of an offering that can't be perceived by the senses. Goodwill, brand, image and position are examples of intangibles. • The role of intangibles in the customer decision process is often underestimated leading to products that are perceived by the customers as incomplete.
Intellectual Property (IP)	• Information, formulas, designs and other knowledge that is unique to an individual or company and can be used to grow a business or improve the competitiveness of an organization. • IP takes many forms, but the main criterion is that it can provide a competitive advantage or can be sold to someone who values it.

Appendix B

Glossary

Term	Definition & Notes
Issue	• A known obstacle, problem or anything that has 100% probability of occurring. And when it does occur, it will cause serious or fatal disruption to a program or project.
Kaizen	• A Japanese term that represents the philosophy of continuous improvement. • More than a set of tools or processes, Kaizen is a state or mind, a commitment to continually improve and striving for perfection.
Kano, Kano Analysis	• A model developed by Dr. Noriaki Kano to characterize quality as perceived by the customer. • Generally shown as a two-dimension diagram, the model maps the degree to which the new product performs versus the degree of satisfaction of the customer. • The model assumes that all input shown was provided by the customers, and not internal personnel. • The model can be very tedious and involved.
Lagging Metrics	• See "Outcome Metrics."
Lead Customers	• Carefully selected representatives of customer groupings (segments) that have needs and requirements typical of the rest of the segment. • In some companies, Lead Customers represent customers who are facing problems ahead of the rest of the segment, providing early visibility of future customer needs.
Leading Metrics	• See "Process Metrics."
Lessons Learned	• The benefit of performing retrospects (postmortems) and implementing the improvements as a result of the process. • Lessons learned do not benefit the organization until they are applied to improve the future performance of the organization.
Life Cycle Costs	• The total costs that the customer will incur during the useful life of the new product. These costs include purchase, maintenance, repair, installations, training operations and disposal. • Generally used during the selling cycle. Customer wants to know the total cost of a new product, not just the selling price.

Market Segment	• A group of customers who display similar characteristics relevant to a business.
	• Market segmentation should answer three questions:
	  – Who are the customers?
	  – Where are they located?
	  – What are they trying to achieve?

Market Segmentation	• The process of grouping customers with similar characteristics. • Generally the segments represent the view of the customers; that is, how they classify themselves. • A market-driven segmentation should include your products or service. It should focus on how customers group themselves. Subsequent versions of the segmentation may include your products.

Glossary

Term	Definition & Notes
Market Sensing	• Research, observations and experience develop a deep understanding of the forces acting on a market to enable the development of a vision of the future. • Sensing the market is generally the responsibility of the senior staff and involves deep participation in the affairs of a market place. • Gathering customer needs and customer requirements is not part of Market Sensing.
Market Share	• A company's sales in a product area expressed as a percent of the total sales into that market by all competitors • Market share depends on what the company defines as its market and its competitors. • If it only considers competitors as companies that offer the same product type, it will have a higher market share than if it defines its competition as including substitution products.
Market-Driven	• A commitment to serve the needs of a market. • A chosen strategy and cultural shift that enable the achievement of a set of business objectives by focusing on serving specific market segments, gaining profound insight into specific market segments, delivering superior value and earning customer loyalty.
Marketing Requirements Document (MRD)	• The process that the marketing function uses to communicate to engineering (R&D) the problems that a new product must address. This takes the form of a document called the MRD, which articulates the prioritized customer requirements. Engineering uses the MRD to generate product concepts, features and specifications • The MRD goes by various names depending on the company or industry. There is always a document that serves to communicate the prioritized customer requirements to the R&D organization.
Mean Time Between Failures (MTBF)	• The number of failures in a given period of time divided by the total uptime in the same period. • A measure of reliability that represents the average time between failures.
Mean Time to Repair (MTTR)	• The total estimated downtime of the new product for a given period divided by the number of failures or repairs performed in the same period. • Used to plan maintenance programs, calculate Total Cost of Ownership and compare reliability.

Metrics	• A standard of measurement. For example, metrics will directly measure our progress toward achieving the organization mission and group's objectives.
	• There are three kinds of metrics:
	  1. Results metrics.
	  2. Process metrics.
	  3. Customer satisfaction metrics.

Appendix B

Glossary

Term	Definition & Notes

Mission	• The specific task or assignment adopted by an entire group or team. The Mission describes what it does, for whom it does it and the reason it does it.
	• The mission statement should be specific, achievable, easy to articulate and motivating. For example:
	  – The mission of the police is to protect and to serve.
	  – The mission of the product team is to get a specified product out on time and meet customer requirements.

Need	• Describes what the customer says he is lacking, what he would like to do or obtain or what he wants to get. • Customer needs, as expressed by customers, are often limited to product "features" (attributes) or problems that may or may not be related to the customer end goals or to how the decision will actually be made between competing offerings. It is the reason we introduced the term "Compelling Reason to Act."
Net Present Value	• Discounting future cash outflows and inflows using today's "cost of money." • A way to normalize several investments so you can compare apples to apples. • A way to determine the viability of an investment objectively. • Used to evaluate several alternative new products and compare their financial viability.
New Product Development (NPD)	• The processes, models and tools to develop a new product. • NPD ensures speed, repeatability and concurrency. • The processes must be documented, measured and improved in order to yield long-term benefits.
New Product Introduction (NPI)	• The process of rolling out or launching a new product to the market. • The go-to-market strategy and its corresponding tactics.
Objective	• A specific, definable and measurable end-result, which is to be pursued by a team or an individual of a company within a specific period of time. • Objective and Goal are often used interchangeably. See this definition.
Offering	• A set of products, services and intangibles offered to satisfy particular customer problems. • Express what the product is from the customer's perspective. • See also "Complete Product" and "Whole Product."
Outbound Marketing	• The function that leads all activities that take the products, programs and messages to the market. • Includes managing the distribution channels, enabling the direct sales channel in all its forms, marketing communications and customer retention programs, pricing and pricing schedules, new product roll-outs, etc.
Outcome	• The measurable outputs of an action project or program. The measurement of the actual results. • It is important not to confuse "completion" of a task with "results"; they are not necessarily the same.

Glossary

Term	Definition & Notes
Outcome Metrics	• Also called "Lagging Metrics," they track the final outcome or result of a project or process.
Pipeline	• See "Roadmap."

Plan	• A detailed formulation of a program of action; for example, a Marketing Plan.
	• A written set of tasks, schedules and resources with defined end dates and an assessment of any barriers which might put the schedule at risk.
	• There are many outlines for a plan, but generally a plan answers seven critical questions:
	  1. What needs to be done?
	  2. Who will do it?
	  3. By when will it be done?
	  4. How much will it cost?
	  5. What are the threats and risks?
	  6. Why are we doing it?
	  7. How will we mange change?

Policy	• A definite course or method of action to guide and determine present and future decisions. • An envelope for action with clearly defined boundaries. • The organization may choose to make policy inviolate to provide a more robust framework for action.
Portfolio Management	• A process to prioritize development projects based on a set of documented criteria. • Can be used any place where a business decision must be made. • The intent is to maximize the value of the portfolio to the corporation and to ensure that corporate strategies are in full alignment with the flow of new products. • Other names include Pipeline Management, Product Pipeline, The Product Funnels and The Product Portfolio.
Postmortem	• A process to understand the effectiveness of a project and to use the information to improve future performance. • Postmortems can be done at any time during the life of a project; they are most commonly done at the conclusion. • Postmortems are only useful if the information is used to improve the future performance of the organizations. • Postmortems are also known as Debriefs, Retrospects and Lessons Learned.
Procedure	• A written series of steps followed in a regular manner to accomplish a specific task in reproducible manner. • Similar to a process, but generally assigned to a single individual. Some compare a task with an action item.
Process	• A documented, sequential list of value-added steps that can replicate a product or service to a customer. • Unlike a project, processes are designed to be repeatable. A project is a single event with a clear start and end.

Appendix B

Glossary

Term	Definition & Notes
Process Metrics	• Also called "Leading Metrics," they measure activities that lead to a result. They are important to track progress towards a goal. These metrics are helpful to improve processes.
Process Out of Control	• When the number of defects exceeds boundary conditions. • Thus, an engineering organization, as a process, is out of control when there are no metrics to detect the number of defects nor a system for remedial and corrective action.
Product	• Anything that can be offered to a market that represents value; i.e., demand. • Good or service that meets the requirements of a market segment. • Colloquially, a manufactured good or the tangible part of value. But this is a limiting connotation, as customers in a business-to-business environment buy solutions to problems, and such solutions are made up of tangibles, services, experiences and intangibles. Sometimes the term "whole product" is used when referring to the entire package of tangibles, services, experiences and intangibles offered to a market.
Product Benefit	• The positive outcomes or results that an offering promises a customer. • Anticipated benefits apply before the purchase. Actual benefits apply after the purchase and manifest themselves as the customer's experience with your offering.
Product Development Team	• All the individuals whose contribution will result in the delivery of a new product to the marketplace. • In some organizations the Product Development Team is defined more narrowly; for example, only the marketing and engineering professionals involved.
Product Funnel	• See "Pipeline."
Product Life Cycle	• The four stages that all new products go through from introduction to end-of-life: introduction, growth, maturity and decline. • Duration of time from the earliest proposal to removal from the order entry system (phase out). A series of phases products follow from earliest conception to end-of-life • Product Life Cycle has many applications in product companies; such as tracking the product portfolio, ensuring profitability (gross margin time number of units), resource allocation, etc.

Glossary

Term	Definition & Notes
Product Management	• The process and strategy to mange the life cycle of a product such that it meets profitability targets. • The role of Product Management is to define, develop, deploy and maintain products and services that: 　– Provide more value than the competition. 　– Help build a sustainable competitive advantage. 　– Deliver financial benefit to the business. • Six of the key disciplines of product management are: 　1. Market Research 　2. Product Definition and Design 　3. Project Management 　4. Evangelize the Product 　5. Product Marketing 　6. Product Life Cycle Management
Product Manager	• The person assigned the responsibility and accountability for leading all the various activities that concern a particular product along its life cycle. • Sometimes "Brand Manager", particularly in consumer products. • Generally the Product Manager has P&L (Profit and Loss) responsibility for the product line. Some companies do not give P&L responsibility, but assign a "pro-forma" P&L to signify the importance of the job. • The Product Manager is responsible and accountable for achieving the business objectives of the product line.
Product Portfolio	• See "Pipeline."
Product Requirements Document (PRD)	• See "Marketing Requirements Document."
Program Evaluation Review Technique (PERT)	• A form of network analysis used to estimate the duration of a project.
Program Management	• The business function that assures the clean execution of corporate strategies, where a program is a group of related projects managed in a coordinated way to obtain the benefits and control not available from managing them individually. • The investment in program management is made to: 　– Reduce the variability of return from an investment. 　– Ensure alignment between strategy and execution in order to meet the business objectives of the enterprise. 　– Deal with change in a dynamic environment to ensure business results. • Program Management is not the same as "Project Management."
Program Manager	• A person with the skills, knowledge and abilities to lead a cross-functional team in the development of a product with a successful outcome such as profitable time-to-market. • This definition applies to the context of product development — your organization may use this title differently within a different context.

Appendix B

Glossary

Term	Definition & Notes
Programmatic Customer Visits	• A structured set of customer interviews where the participants share a common set of topics to be discussed. Further, the participants agree to a regular schedule of trips to the field to conduct the interviews. • A common way to capture, prioritize and archive the information from the customers.
Qualify Function Deployment (QFD)	• Quality Function Deployment is a structured methodology that uses a set of matrices for linking market requirements to the entire product development effort. • QFD will not define a product for you; it is simply a tool to manage information about the new product so that the integrity of the customer requirements is not jeopardized. • QFD is also called the House of Quality due to the appearance of the matrices. • QFD originated in Japan as a way to manage all the features and their interactions of a new product from start to finish.
Quality	• Perception of the degree to which a product or service meets customers' expectations. • Quality is whatever your customer says it is. • Quality is an attribute of the product that generates customer satisfaction. But it is also a perceived attribute, and therefore is somewhat subjective. • Quality Standards are approved, documented rules that define one or several dimensions of quality for a particular application or set of customers.
Rapid Prototyping	• Any of a variety of processes that avoid tooling time in producing prototypes. Generally these are nonfunctioning. • Rapid Prototyping exists to test the feasibility of a particular design with customers to reduce time-to-market.
Remedial Action	• Activities to provide an immediate resolution in order to solve a problem to the satisfaction of the customer. • Without regard to permanent solutions.
Request for Proposal (RFP)	• A process used by larger companies, government entities, military and universities to invite suppliers to bid on a specific acquisition of products, services, maintenance, etc. as a package. The submitted RFPs are then used by the buying entity to evaluate the merits and make an informed decision to purchase.
Request for Quotation (RFQ)	• See "Request for Proposal (RFP)."
Requirement (Customer)	• A brief and clear statement expressing what the customer will be able to experience as a result of owning and/or using the offering. • Product and customer requirements are two ways to express the same thing. • Requirements must be short, unambiguous, and able to be verified
Requirement (Product)	• A brief and clear statement expressing what an offering will have such that the needs of the customer will be met. • Product and customer requirements are two ways to express the same thing. • Requirements must be brief, unambiguous, and able to be verified.

Glossary

Term	Definition & Notes
Results	• The measurable achievement of an endeavor; for example, the results of a winning product is the achievement of the profitability target. • The outcome of an endeavor.
Rev (Review or Revise)	• To review or revise, used in software development to track the number of revisions that have been made to a specific application. • Revision can also be applied to documents to track version control.
Reverse Engineering	• The process of copying a product by analyzing every part of a competitor's product and then incorporating its features into one's own products. • Generally not a good thing to do.
Risk	• Something that can harm your program and should not happen. • The probability that something may go wrong with your project. • There are many types of risks that can interfere with your product development.
Risk Management	• The process to identify, prioritize, plan and manage risks in a project. • Risk Management starts at the earliest stages of a project and continues through completion.
Risk Management Plan	• The documented set of prioritized risks and corresponding mitigation plan. • The Risk Management Plan is constantly monitored.
Roadmap	• The manifestation of the product strategy for a product line or a company. • The outline of the future products versus the time it takes the organization to develop to achieve its business objectives. • Organizations may use terms like "Funnel", "Product Funnel", "Product Roadmap", "Pipeline" or "Product Portfolio." We suggest all of these terms are synonymous.
Simultaneous Engineering	• See "Concurrent Engineering"
Situation Analysis	• The current snapshot of the factors confronting a team, company or organization in a complex situation. • A short list of items, observations or data that are highly relevant to the topic at hand and are presently important to consider. • Examples include the Situation Analysis of a market and the Situation Analysis in a Product Proposal.
Sponsor, Team Sponsor. Team Mentor	• An Executive of the company who agrees to support a team in order to ensure its success. • The sponsor must have the time, willingness and skills to assist the team in a measurable way.
Strategic Alignment	• The endeavor to align all stakeholders on a single strategic direction such that their day-to-day activities fully support the endpoint. • Strategic Alignment complements (but is different from) "Strategy Formulation" or "Strategizing", which is the process of synthesizing and articulating a particular strategy.

Appendix B

Glossary

Term	Definition & Notes
Strategic Formulation (Strategizing)	• The act of formulating a strategy, different from the process of strategic planning. • Strategizing is the thought process or brainstorming needed to synthesize a strategy given a clear understanding of the market.

Strategic Initiative	• A program that, when completed, will deliver visible results in clear support of an overall strategy. It's defined to clearly move the organization towards the achievement of the corporate or group strategic goal.
	• Strategic Initiatives possess the following attributes:
	   – Deemed critical to the success of the organization — non-negotiable and successful implementation is the only outcome.
	   – In general, assigned to a member of the executive team as part of his MBO.
	   – In general, implementation is carried out by one or more small teams that are given a clear charter, including deliverables, metrics, desired outcomes, time lines, roles and responsibilities. The executive is still accountable for measurable results.
	   – Periodic reviews are included to ensure progress and accountability.
	   – Fulfilled through programs and projects.
	   – Span more than one fiscal year or are part of the yearly operational plan.

Strategy	• The approach that a team or company chooses to achieve a set of business objectives. • The approach guides the choices, behavior and direction of an organization.
Strategy (Corporate)	• A set of Strategic Initiatives made up of programs and projects that, when implemented, will fulfill the vision of the company. • This is the means by which you connect everyone's day-to-day activities to the vision of the company.
Strategy (Year Around Planning)	• A model of strategic planning that, unlike the yearly cycle of "business planning", assumes that strategy is a constant endeavor. • Usually characterized by a rigorous review and update of the strategy twice or three times per year.
Superior Value	• Better than your competition. Always defined by your customers, based on their comparison of you with your competitors. • Superior Value shows when a customer selects your products or solutions over that of your competition.
Switch-Over-Cost	• A term used to describe the mental process to shift one's cognitive resources from one topic to a completely different one. • Research done by Pattison, Kermit and Worker in "The Cost of Task Switching" shows that it may take up to 23 minutes to return to the same level of concentration after an interruption.
Team Charter	• See "Charter."
Team Sponsor, Team Mentor	• See "Sponsor."

Glossary

Term	Definition & Notes
Theory of Inventive Problem Solving (TRIZ)	• A Russian acronym that refers to a methodology for solving a difficult problem whereby multiple alternative solutions are defined. • The method is supposed to improve creativity by enabling the team to think of many solutions to a single problem. Generally a multi-discipline methodology to tap of the creativity of different individuals.
Timeline	• The schedule or the amount of time plus the activities needed to deliver a set of results. • A Gantt chart showing the chronological progression to the completion of a project.
Time-to-Market	• A measure of the speed by which a new product is developed. • Measured from the official start of the development until the product is ready to be purchased by a customer. • A critical measure to ensure competitiveness and success in the market place. • Various companies use different start and end points to calculate their time-to-market.
Time-To-Results (Time-To-Profit or Time-To-Cash)	• That point in the product life cycle where a target result that determines the success of the product has been met. • The success of a product may be measured by margin, revenue, market share or other strategic goals. • The time-to-results point is determined no later than the creation of the product proposal.
Total Cost of Ownership (TCO)	• The aggregate of all the costs during the life of a product, from acquisition to disposal.
Use Case	• Describes a specific activity or set of activities performed by a customer with a device or product. • Use cases are created with specific goals to understand the interaction between an operator and the equipment.
Value	• The customer's perception of what he wants to happen in a specific use and situation as the result of using a company's offering. • There is value in using the product or service — i.e., deriving a benefit as a result of use. There is also value in owning the product or service — i.e., deriving an emotional benefit just from owning the product or service.
Value Proposition	• A statement describing what resulting experience the customer will have, articulated in a way that communicates clearly and briefly the benefits, costs and competitive advantage of a particular offering. • The value proposition must be compelling and differentiated as perceived by the customer. • The resulting experience includes the outcomes or results that the customer will experience. • Whole products deliver value propositions.

Appendix B

Glossary

Term	Definition & Notes
Values	• A set of deeply held beliefs that enable employees to act without the need to seek management approval. • Values are never intentionally compromised. • A common set of social principles that guide the behavior and decisions of each person and the organization.
Vision	• The vivid description of the end destination and results of a group effort. • The distant lighthouse that provides direction to the completion of the team's journey. • A vision statement evokes concrete images. It is compelling and prompts forward progress. It is succinct and memorable.
Voice of the Customer (VOC)	• A process that uses structured, in-depth interviews for eliciting customer requirements. • VOC is also used as a generic way to talk about customer needs or customer requirements. • VOC by itself does not yield customer needs; other processes have to be used in conjunction to process the input from the customer, turn it into product features and eventually into a new product.
Voices	• Verbatim quotes from customers that provide insights as to their needs. • These voices are processed to extract requirements.
Whole Product	• The set of product, services, ancillary parts and intangibles needed to ensure that the customer will achieve his compelling reason to act (CRTA). • The actual product is a subset of the whole product. The whole product is what the customers buy. • Ignoring dimensions of the whole product explains why so many companies fail to penetrate new markets. • See also "Complete Product" and "Offering."

Appendixes

APPENDIX C

Templates

Instructions

These templates are for you to use and share. Throughout this guidebook, you have seen real world tools and examples. Any page with a template icon indicates that there is a correlating fillable form in this section.

For more resources and future updates, visit us at: www.rapidinnovation.com.

Templates with Examples

Example: Selecting the Right Project to Debrief	20
Example: A Well-Documented Action Plan	32
Example: Debrief Planning Tool	33
Example: Debrief Planning Tool by Focus (Topic)	36
Planning Checklist	44
Planning Document	45

More Tips for Users of the Interactive PDF Version

You can view the complete list of templates with the "Show/Hide Templates" bookmark on the left Navigation Panel. This reveals the Attachments Tab at the bottom of the screen. Double click any template you want to work on. You can also open a template by double clicking the white pushpin icon on the specially marked example pages.

Once opened, these templates are enabled for use in Adobe Reader or Adobe Acrobat — templates that you can fill with your own information and share with your peers and colleagues.

When you open original templates from your guidebook, they will be fresh. But you can save versions for future access — whether you have completed the template, partially filled in your information or left in blank. Saved versions remain enabled for revisions and collaboration.

When you close the template, you will return to this guidebook.

These enabled templates also provide you with Comment & Mark-up Tools. Just as you might with a hard copy, you and your team can highlight text, add notes, and make custom annotations.

Debriefs and Postmortems for Product Development
Do it now. Do it right. Deliver results. Templates for the Real World.

Selecting the Right Project to Debrief

Use this simple tool to select the right project. Develop the template to track, prioritize and optimize the results from debriefs.

Project Candidate	Business Case	Focus (Topic)	Plan Leader	Target Date	Deployment Leader
Selection of projects is based on a clear criteria, which is applied to all candidates.	The business rationale that justifies the expense in resources to perform the debrief.	The specific area of focus for the debrief. The desired outcome will dictate the focus.	The name of the person responsible for leading the planning and execution of the debrief.	The intended date when the main session of the debrief is planned.	The name of the person who is responsible to lead the implementation of the results of the debrief.

For more resources and future updates, please visit us at: www.rapidinnovation.com.
Contact the author: debriefs@rapidinnovation.com. ©1992, 2012 Rapidinnovation, LLC.

RealWorld Product Development Series

Debriefs and Postmortems for Product Development

Do it now. Do it right. Deliver results. Templates for the Real World.

A Well-Documented Action Plan from a Debrief

Item	Background Info	Action Item	Responsibility	Timeline

For more resources and future updates, please visit us at: www.rapidinnovation.com.
Contact the author: debriefs@rapidinnovation.com.   ©1992, 2012 Rapidinnovation, LLC.

RealWorld Product Development Series

Debriefs and Postmortems for Product Development

Do it now. Do it right. Deliver results. Templates for the Real World.

Debrief Planning Tool

To be filled out by the person responsible for organizing the upcoming debrief.

Item	Description

For more resources and future updates, please visit us at: www.rapidinnovation.com.
Contact the author: debriefs@rapidinnovation.com. ©1992, 2012 Rapidinnovation, LLC.

RealWorld Product Development Series

Debriefs and Postmortems for Product Development

Do it now. Do it right. Deliver results. Templates for the Real World.

Debrief Planning Tool by Focus (Topic)

This is a comprehensive form to enable conversations. It may be too complex for some projects or programs, so we suggest that you customize this form. Do not send this form to team members! This form is to help you facilitate the meeting, but the real value is the dialog.

1. Planning

2. Resources

3. Scheduling

4. Performance of Project or Program Management

5. R&D / Technology Management

6. Communications

For more resources and future updates, please visit us at: www.rapidinnovation.com.
Contact the author: debriefs@rapidinnovation.com. ©1992, 2012 Rapidinnovation, LLC.

RealWorld Product Development Series

Debriefs and Postmortems for Product Development

Do it now. Do it right. Deliver results. Templates for the Real World.

Debrief Planning Tool by Focus (Topic) cont'd

7. Managers and Executives

8. Product

9. Marketing

10. Manufacturing (In Software Development, use Documentation or QA)

11. Virtual or Distributed

12. Service and Repair

13. General Questions

Debriefs and Postmortems for Product Development

Do it now. Do it right. Deliver results. Templates for the Real World.

Planning Checklist for Deploying a New Tool

Deploying a tool is indeed a project. Consequently, ensure that you have a deployment plan, a schedule, budget, etc. Working with your Core Team, or with the peers who have agreed to help, you can develop a tactical plan that will ensure a successful deployment.

The list below is intended to help you formulate your own plan. You may need to add your own items to the list. Also note that the list below is in no particular sequence. The sequence will be determined by your specific needs and schedule. Finally, the list is not rigid. Use you judgment and do what makes sense; you know your organization better than most.

Planning Checklist

- ❑ Schedule for the deployment defined, documented and approved.
- ❑ Training materials for employees developed and documented.
- ❑ Training materials produced (hard & soft copies), available for use.
- ❑ Training materials placed in the intranet or Web page.
- ❑ Trainers on the new tool or method identified and informed of their duties.
- ❑ Training of employees scheduled and employees notified.
- ❑ Training of managers and supervisors scheduled.
- ❑ Feedback mechanism in place.
- ❑ Metrics to measure the progress and eventual success of the deployment identified and ready to use.
- ❑ Coaching and hand-holding for managers, employees and stakeholders defined and ready — including those who will do the coaching.
- ❑ Support (assistance) available; for example, facilitators, coaches, tech support, etc.
- ❑ Obstacle removal in place.
- ❑ Templates, checklists and other tools ready and available.
- ❑ Information on our Web page done.
- ❑ Processes that will be impacted by the deployment identified and corrected.
- ❑ Changes to our quality manual completed.
- ❑ Manuals, guides and other reference materials that will be needed by the users ready and available.
- ❑ Executives briefed and ready to fully support the deployment.
- ❑ Human resources (HR) informed.
- ❑ Training department informed.

For more resources and future updates, please visit us at: www.rapidinnovation.com.
Contact the author: debriefs@rapidinnovation.com. ©1992, 2012 Rapidinnovation, LLC.

RealWorld
Product Development Series

Debriefs and Postmortems for Product Development

Do it now. Do it right. Deliver results. Templates for the Real World.

Planning Document for Deploying a New Tool

This tool may help you secure assistance from your executives. It is also a good point to start the process of deployment. It may even become a standard template for deployment of future tools and templates. You will want to modify it to meet your specific objectives and adapt to the culture of your organization.

1. Tool or methodology to be deployed

2. Objective of the tool deployment

3. Who is the "expert" on this tool?

4. What documentation is available?

5. Are there any external resources available?

6. What are the risks involved in the deployment?

7. How will we "pilot" the tool?

8. What will be our communication plan?

9. What will we measure to assure us that the deployment has been successful?

For more resources and future updates, please visit us at: www.rapidinnovation.com.
Contact the author: debriefs@rapidinnovation.com. ©1992, 2012 Rapidinnovation, LLC.

RealWorld
Product Development Series

How to Get Your Interactive Version of this Guidebook

FREE BONUS

Register Your Book

Thank you for purchasing *Debriefs and Postmortems for Product Development*. Many find the interactive PDF version of this guidebook to be a valuable addition because the cross-reference links are live and all the templates can be filled out electronically and used to collaborate with your team members around the world. Once you enter the information, every template becomes a separate PDF file for sharing and collaboration. Furthermore, the electronic version of this guidebook can also be stored in any portable device that accepts PDF files such as tablets, laptops and even smartphones.

To obtain your free interactive PDF copy of *Debriefs and Postmortems for Product Development*, follow these simple steps:

1. Send an email to debriefs@rapidinnovation.com with "**Registering DEBRIEFS book**" in the subject line. In the body of your email, include your name plus the following registration code: **DBF4818E**.

2. In a few days, you will receive an email notice that your book has been registered with either an attached copy of your free interactive PDF or a link to download your copy.

If you have any questions during registration, please email the publisher at: debriefs@rapidinnovation.com.

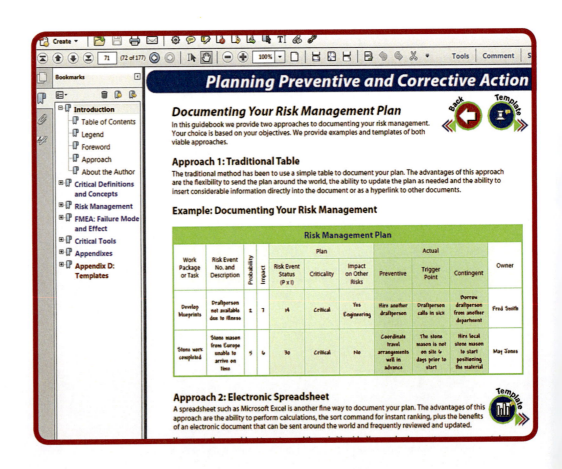

Other Titles You May Be Interested in Owning

THE SERIES

Risk Management and FMEA for Product Development

"If you are not managing risk, you are not managing your project." This is most applicable when it comes to managing risk in product development. Risk is the probability of something going wrong with your development project. Technology changes, customer requirements vary, schedules are always in flux and your suppliers are in turmoil. In short, many things that can go wrong. But what if you could predict what could go wrong? And more importantly, what if you could develop a plan to eliminate or mitigate your risks?

Risk Management is the ability to anticipate risks and develop plans to manage them; thus, creating more predictable schedules, improving reliability and customer satisfaction. In many cases, risk management prevents what otherwise might be a catastrophic failure of your development program.

This guidebook was written by product developers with many years of experience in the "real world," where traditional text book approaches fail. This is a how-to guide that understands the pressures placed on you, the need to get to the point quickly and to show you step-by-step how to apply risk management on product development projects.

The Voice of the Customer for Product Development

Your task is to create differentiated value for your customer in order to achieve the right gross margin at minimum COGS. This guidebook is based on experiences with many technology development teams around the world — all of them focused on achieving the right volume, margin and profitability. You will be introduced to a customer-driven methodology that has been perfected over the last 30 years.

Our mission is to create awareness of the framework and motivation needed to successfully understand customer needs, process the information and put it to use to develop differentiated products that achieve the right level of profit. This guidebook methodically walks you through the tools, templates and methods for obtaining, processing and using customer input — like no other of its kind, it's based on real projects and is documented to provide maximum guidance to enable application.

You will be able to increase the value of your offering offerings and create defensible market segments. You will learn how to identify opportunities by interviewing the right customers. You will also learn how to transform these opportunities into requirements and then into product concepts that cast an optimum balance between usefulness, usability and desirability. The overall objective is to create the capacity for improved productivity!

How to Order
Inquire directly from the publisher at:
info@rapidinnovation.com.
Visit us at:
www.rapidinnovation.com.

Flexible Project Management for Product Development

Traditional project management approaches are not sufficient for product development due to the complexity of getting a new product to market. Technology changes, personnel turnover, changing customer requirements, multi-site development, and unclear schedules are but a few of the variables that render conventional project management deficient.

Flexible Project Management for Product Development allows you and your team to manage your development project in the middle of the turmoil created by global competition. Beyond just reading very useful material, you will be able to use the templates and tools and also be able to share the information with your peers around the world. This is a step-by-step how-to guide. The authors realize that in the real world, you need more guidance and enough detail to enable you to apply this material to your specific development project. It starts by helping you to organize your development project to ensure that you build a robust environment to deal with chaos.

Made in the USA
San Bernardino, CA
24 July 2017